NEW CONCISE PROJECT MATHS 3

FOR LEAVING CERT ORDINARY LEVEL

GEORGE HUMPHREY, BRENDAN GUILDEA, GEOFFREY REEVES

GILL & MACMILLAN

Gill & Macmillan
Hume Avenue
Park West
Dublin 12
with associated companies throughout the world
www.gillmacmillan.ie

© George Humphrey, Brendan Guildea and Geoffrey Reeves 2011

978 07171 5033 5

Print origination by MPS Limited, a Macmillan Company

The paper used in this book is made from the wood pulp of managed forests. For every tree felled, at least one tree is planted, thereby renewing natural resources.

All rights reserved.
No part of this publication may be copied, reproduced or transmitted in any form or by any means without written permission of the publishers or else under the terms of any licence permitting limited copying issued by the Irish Copyright Licensing Agency.

Any links to external websites should not be construed as an endorsement by Gill & Macmillan of the content or views of the linked materials.

Contents

Preface — iv

1. Algebra — 1
2. Arithmetic — 50
3. Complex Numbers — 99
4. Perimeter, Area and Volume — 127
5. Patterns, Sequences and Series — 193
6. Functions — 228
7. Graphing Functions — 237
8. Differentiation — 262

Answers — 286

Acknowledgments — 308

Preface

New Concise Project Maths 3 (Strands 3, 4 & 5) covers the Strand 3 and Strand 4 course changes for Leaving Certificate Ordinary Level students who will take the exam in 2013. It includes the existing Strand 5 material, which does not change for students taking the exam in 2013.

New Concise Project Maths 3 (Strands 3, 4 & 5) incorporates the approach to the teaching of mathematics envisaged in **Project Maths**. It reflects the greater emphasis on the understanding of mathematical concepts, developing problem-solving skills and relating mathematics to everyday events.

The authors strongly empathise with the main aims and objectives of the new Project Maths syllabus and examination. In the worked examples, a numbered, step-by-step approach is used throughout the book to help with problem solving. There is a comprehensive range of carefully graded exercises to reflect the new exam. Exam-style in-context questions are included to enhance students' understanding of everyday practical applications of mathematics. The emphasis is on a clear and practical presentation of the material. Simple and concise language is used throughout, instead of technical language, which is not required in the exam.

George Humphrey
Brendan Guildea
Geoffrey Reeves
May 2011

CHAPTER 1

ALGEBRA

Evaluating expressions (substitution)

A **substitute** is used to replace something. In football, a substitute replaces another player. In algebra, when we replace letters with numbers when evaluating expressions, we call it **substitution**. When you are substituting numbers in an expression, it is good practice to put a bracket around the number that replaces the letter. (Remember: **BEMDAS**.)

EXAMPLE

Find the value of the following.

(i) $4(a-b)^a$ when $a=2$ and $b=-1$

(ii) $\dfrac{5(x-y)}{3(x^2+y^2)}$ when $x=2$ and $y=-1$

Solution:

(i) $4(a-b)^a$
$= 4[(2)-(-1)]^2$
$= 4[2+1]^2$
$= 4(3)^2$
$= 4(9)$
$= 36$

(ii) $\dfrac{5(x-y)}{3(x^2+y^2)}$
$= \dfrac{5[(2)-(-1)]}{3[(2)^2+(-1)^2]}$
$= \dfrac{5(2+1)}{3(4+1)}$
$= \dfrac{5(3)}{3(5)} = \dfrac{15}{15} = 1$

Exercise 1.1
Evaluate each of the following in questions 1–22.

1. $3 + 4 \times 5$
2. $-3 + 7$
3. $-5 + 2$
4. $-1 - 5$
5. $8 - 10$
6. $5(-2)$
7. $-3(4)$
8. $-2(-4)$
9. $2(3)^2$
10. $(-4)^2$
11. $5(-2)^2$
12. $-4(-5)^2$
13. $-20 \div 5$
14. $12 \div -6$
15. $-18 \div -3$
16. $-8 \div 8$
17. $(5)^2 - 2(5)$
18. $(4)^2 - 2(4) - 8$
19. $(-3)^2 - 4(-3) - 21$
20. $\dfrac{6+2}{6-2}$
21. $\dfrac{18 + 2 \times 3}{2(5-1)}$
22. $\dfrac{5^2 - 7}{3^2 - 7}$

Evaluate each of the expressions in questions 23–32 for the given values of the variables.

23. $x^2 + 2x + 3$, when $x = 1$
24. $x^2 - 3x + 2$, when $x = -2$
25. $a^2 + 2\sqrt{a} - 20$ when $a = 4$
26. $x^2 - 5xy$ when $x = 3$ and $y = -2$
27. $3(2p - q)$ when $p = -4$ and $q = 5$
28. $(a - b)^a$ when $a = 2$ and $b = -1$
29. $\dfrac{3x - 2y - 1}{5}$ when $x = 13$ and $y = 14$
30. $\dfrac{p^2 + 4q}{2(q + 1)}$ when $p = -2$ and $q = 3$
31. $\sqrt{3a - 2b}$ when $a = 4$ and $b = -2$
32. $(3x^2 - 11)^{\frac{1}{2}}$ when $x = 5$

33. Find the value of $5x - 3y$ when $x = \dfrac{5}{2}$ and $y = \dfrac{2}{3}$.

34. Find the value of $\dfrac{ab - c}{2}$ when $a = 3$, $b = \dfrac{2}{3}$ and $c = 1$.

35. Find the value of $\dfrac{a + b - 1}{a - b + 1}$ when $a = \dfrac{1}{2}$ and $b = \dfrac{2}{3}$.

36. A car increases its speed from u km/h to v km/h in a time t hrs.

    The distance, d km, that it has travelled is given by the formula $d = \dfrac{t}{4}(u + v)$.

    Calculate d when $t = 12$, $u = 3$ and $v = 7$.

37. x people share the cost of travelling y km in z cars.

    The amount, €A, each has to pay is given by the formula $A = \dfrac{z\left(8 + \dfrac{y}{5}\right)}{x}$.

    Find the value of A when $x = 17$, $y = 300$ and $z = 18$.

38. x and y are positive or negative whole numbers such that $2x + y = 8$. Find four pairs of values for x and y that make this equation balance. For example, $x = 5$ and $y = -2$ make the equation balance.

39. $A = 3p + 2q$ where p and q are different positive whole numbers. Choose values for p and q such that A is:

    (i) Even   (ii) Odd   (iii) Divisible by 5   (iv) A perfect square   (v) A prime number

40. x and y are positive whole numbers. Explain why $(6x + 4y)$ is always even.

41. Patrick knows that a, b and c have values 5, 6 and 10 but he does not know which variable has which value.

    (i) What is the maximum value that the expression $3a + 2b - 5c$ could have?

    (ii) What is the minimum value that the expression $5a - b + 2c$ could have?

42. The number x is a positive whole number. Write, in terms of x, the next two positive whole numbers greater than x. Show that the sum of these three numbers is always a multiple of 3.

ALGEBRA

Simplifying algebraic expressions

Only terms that are the same can be added.

EXAMPLE

Simplify (i) $2(a^2 + 3a) - a(2a + 5) + a$ (ii) $(2x + 3)(x^2 - 5x - 4)$

Solution:

(i) $\quad 2(a^2 + 3a) - a(2a + 5) + a$
$= 2a^2 + 6a - 2a^2 - 5a + a$
$= 2a^2 - 2a^2 + 6a + a - 5a$
$= 2a$

(ii) $\quad (2x + 3)(x^2 - 5x - 4)$
$= 2x(x^2 - 5x - 4) + 3(x^2 - 5x - 4)$
$= 2x^3 - 10x^2 - 8x + 3x^2 - 15x - 12$
$= 2x^3 - 10x^2 + 3x^2 - 8x - 15x - 12$
$= 2x^3 - 7x^2 - 23x - 12$

Exercise 1.2

Simplify each of the following in questions 1–11.

1. $4x + 3x$
2. $8x - 6x$
3. $5x - 8x$
4. $-2x - 4x$
5. $-x + 5x$
6. $2a - 9a$
7. $-8y - 6y$
8. $-4b + 7b$
9. $2x^2 + 3x^2 + 4x^2$
10. $-2a^2 + 5a^2 - a^2$
11. $-x^2 + 3x^2 + 5x^2 - 4x^2$

Multiply these terms in questions 12–19.

12. $(2x)(3x)$
13. $(-2x)(5x)$
14. $(3x^2)(-4x)$
15. $(-2x^2)(-5x)$
16. $(-x)(-x)$
17. $(-3a)(-4a)$
18. $(3y^2)(-5y)$
19. $(-2p)(-4p^2)$

Expand (remove the brackets) for each of the following in questions 20–26 and simplify.

20. $(x + 2)(x + 3)$
21. $(2x + 5)(x - 4)$
22. $(3x - 2)(2x - 5)$
23. $a(a - b) + b(a - b) + b^2$
24. $2(x^2 + 3x) - x(2x + 5) + x$
25. $a(a + 1) + 2a(a - 3) + 3(2a - a^2)$
26. $a(b + c) - b(c - a) - c(a - b)$

3

27. Match up the following Algebra Snap cards into groups. Which card is the odd one out?

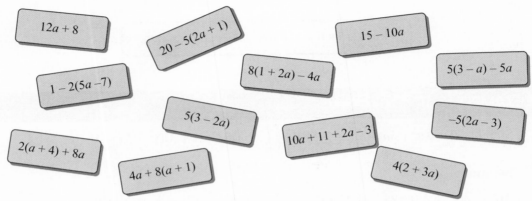

In questions 28–31, show that each of the following reduces to a constant (number) and find that constant.

28. $3(4x + 5) - 2(6x + 4)$
29. $3a(2a + 3) - 6a(a + 2) + 3(a + 1)$
30. $(x - 3)(x + 5) - x(x + 2) + 15$
31. $a(b + c) - b(c + a) - c(a - b)$

32. Write down the simplest possible expression for the perimeter of each shape.

(i) 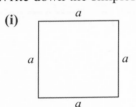   (ii)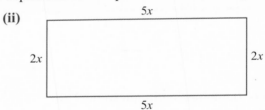

(iii)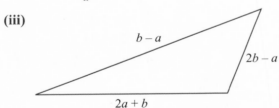

33. Write down the simplest possible expression for the area of each shape.

(i) 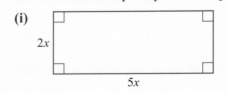   (ii)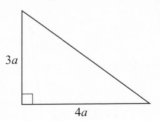

(iii)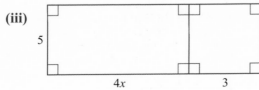

34. The rectangle is made up of 24 squares. Its length is $16\,y$ m. Write down its area is terms of y.

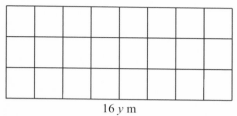

$16\,y$ m

35. In a magic square, the sum of each row, column and diagonal is the same. Which of the following squares are magic squares? In each case, justify your answer.

(i)

$p^2 - 2$	$p^2 - 1$	$p^2 + 3$
$p^2 + 5$	p^2	$p^2 - 5$
$p^2 - 3$	$p^2 + 1$	$p^2 + 2$

(ii)

$a + b$	$a - b - c$	$a + c$
$a - b + c$	a	$a + b - c$
$a - c$	$a + b + c$	$a - b$

(iii)

$y - x$	$3x + 2y$	x
$3x$	$x + y$	$2y - x$
$x + 2y$	x	$3x + y$

36. John wrote the following: $4(3x - 1) + 10(2x + 3) = 7x - 4 + 20x + 30 = 27x - 26$.
    John has made two mistakes in his working. Explain the mistakes that John has made.

Single variable linear equations

A equation is solved with the following method:

> Whatever you do to one side, you must do **exactly the same** to the other side.

Note: Keep balance in mind.

The solution of an equation is the number that makes both sides balance.

EXAMPLE

Solve: (i) $4(x + 5) - 2(x + 3) = 12$

(ii) $\dfrac{x-1}{4} - \dfrac{1}{20} = \dfrac{2x-3}{5}$

Solution:

(i) $4(x + 5) - 2(x + 3) = 12$

$\quad 4x + 20 - 2x - 6 = 12 \quad$ (remove brackets)

$\quad 2x + 14 = 12 \quad$ (simplify the left-hand side)

$\quad 2x = -2 \quad$ (subtract 14 from both sides)

$\quad x = -1 \quad$ (divide both sides by 2)

(ii) The LCM of 4, 5 and 20 is 20. Therefore we multiply each part by 20.

$\dfrac{(x-1)}{4} - \dfrac{(1)}{20} = \dfrac{(2x-3)}{5}\quad$ (put brackets on top)

$\dfrac{20(x-1)}{4} - \dfrac{20(1)}{20} = \dfrac{20(2x-3)}{5}\quad$ (multiply each part by 20)

$5(x-1) - 1 = 4(2x-3) \quad$ (divide the bottom into the top)

$5x - 5 - 1 = 8x - 12 \quad$ (remove the brackets)

$5x - 6 = 8x - 12 \quad$ (simplify the left-hand side)

$5x = 8x - 6 \quad$ (add 6 to both sides)

$-3x = -6 \quad$ (subtract $8x$ from both sides)

$3x = 6 \quad$ (multiply both sides by -1)

$x = 2 \quad$ (divide both sides by 3)

Exercise 1.3

Solve each of the following equations in questions 1–22.

1. $2x = 10$
2. $3x = -12$
3. $-4x = -8$
4. $-5x = 15$
5. $3x - 1 = 11$
6. $7x + 1 = 22$
7. $4x + 7 = -13$
8. $3x - 1 = -10$
9. $5(x + 4) - 3(x - 4) = 40$
10. $10(x + 4) = 3(2x + 5) + 1$
11. $2(7 + x) - 4(x + 3) = 15(x - 1)$
12. $2 - 6(2 - x) = 5(x + 3) - 23$
13. $5 + 2(x - 1) = x + 4(x - 3)$

14. $11 - 2(2x - 5) = 5(2x + 1) - 4(3x - 1)$

15. $\dfrac{x}{2} + \dfrac{x}{3} = \dfrac{5}{6}$

16. $\dfrac{3x}{4} = \dfrac{2x}{3} + \dfrac{5}{12}$

17. $\dfrac{x+2}{3} + \dfrac{x+5}{4} = \dfrac{5}{2}$

18. $\dfrac{x-1}{5} = \dfrac{17}{5} - \dfrac{x+3}{2}$

19. $\dfrac{x}{5} - \dfrac{11}{15} = \dfrac{x-3}{6}$

20. $\dfrac{3x-1}{2} = \dfrac{x}{4} + \dfrac{9}{2}$

21. $\tfrac{1}{3}(4x + 1) - \tfrac{1}{2}(2x + 1) - \tfrac{1}{6} = 0$

22. $\tfrac{5}{6}(3x - 4) - \tfrac{3}{2}(4x + 2) = \tfrac{2}{3}$

23. A and B have the same number of coins.
    A has 2 bags of coins and 23 extra coins.

    B has 3 bags of coins and 6 extra coins.

    Each bag has the same number of coins in it. How many coins are in each bag?

24. Find the value of x in each balance.

    (i)

    (ii)
    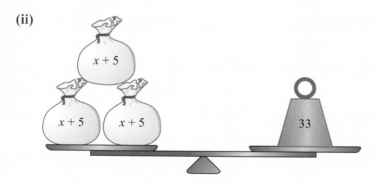

(iii)

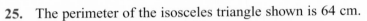

25. The perimeter of the isosceles triangle shown is 64 cm.
    (i) Using this information, write down an equation in terms of x.
    (ii) Solve the equation to find x.
    (iii) What is the length of the base, $(5x - 1)$ cm?

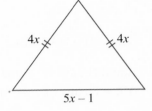

26. Two cylindrical buckets hold 18 litres and 6 litres of liquid, respectively. Another $2x$ litres of liquid are added to each bucket so that the first one now holds twice as much as the second one.
    (i) Express the volume of liquid in each bucket in terms of x.
    (ii) Form an equation in x.
    (iii) Solve the equation to find the value of x.

27. A teacher has a large pile of cards. An expression for the number of cards is $(8n + 10)$. The teacher puts the cards into two piles.
    (i) If one pile has $(3n + 6)$ cards, how many cards are in the other pile?
    (ii) One pile has $(5n + 1)$ cards. If there are 61 cards in this pile, how many cards are in the second pile?

28. Three consecutive numbers (e.g. 3, 4, 5) are x, $x + 1$ and $x + 2$. When the three numbers are added together the result is 33.
    (i) Use this information to form an equation.
    (ii) Solve the equation to find the value of x.
    (iii) What are the three numbers?

29. A girl bought a coat for €$\frac{x}{2}$ and a hat for €$\frac{x}{5}$. The total amount of money she spent was €70.
    (i) Use this information to form an equation.
    (ii) Solve the equation to find the value of x.
    (iii) Find the cost of her hat.

30. A rectangle has sides 8 cm and $(5x + 3)$ cm.
    A smaller rectangle with sides 5 cm and $(2x + 1)$ cm is cut from the larger rectangle. If the remaining area is 139 cm², calculate the value of x.

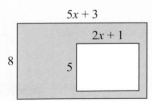

31. (i) What is an equilateral triangle?
    (ii) Could the triangle shown be an equilateral triangle? Justify your answer.

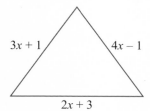

Simultaneous linear equations

Simultaneous linear equations in two variables are solved with the following steps.

1. Write both equations in the form $ax + by = k$, where a, b and k are whole numbers, and label the equations ① and ②.
2. Multiply one or both of the equations by a number in order to make the coefficients of x or y the same, but of opposite sign.
3. Add to remove the variable with equal coefficients but of opposite sign.
4. Solve the resultant equation to find the value of the remaining unknown (x or y).
5. Substitute this value in equation ① or ② to find the value of the other unknown.

EXAMPLE 1

Solve for x and y: $2x + 3y - 8 = 0$ and $\dfrac{3x}{2} + y - 1 = 0$.

Solution:
First write both equations in the form $ax + by = k$ and label the equations ① and ②.

$2x + 3y - 8 = 0$

$\quad 2x + 3y = 8$ ① (in the form $ax + by = k$: label the equation ①)

$\dfrac{3x}{2} + y - 1 = 0$

$3x + 2y - 2 = 0$ (multiply each part by 2)

$\quad 3x + 2y = 2$ ② (in the form $ax + by = k$: label the equation ②)

Now solve between equations ① and ②:

$$2x + 3y = 8 \quad ①$$
$$3x + 2y = 2 \quad ②$$
$$\overline{4x + 6y = 16} \quad ① \times 2$$
$$-9x - 6y = -6 \quad ② \times -3$$
$$\overline{-5x = 10} \quad \text{(add)}$$
$$5x = -10$$
$$x = -2$$

Put $x = -2$ into ① or ②:

$$2x + 3y = 8 \quad ①$$
$$2(-2) + 3y = 8$$
$$-4 + 3y = 8$$
$$3y = 12$$
$$y = 4$$

∴ the solution is $x = -2$ and $y = 4$.

Solution containing fractions

If the solution contains fractions, the substitution can be difficult. In such cases, the following method is useful:

1. Eliminate y and find x.
2. Eliminate x and find y.

EXAMPLE 2

Solve the simultaneous equations $2x + 3y = -2$ and $3x + 7y = -6$.

Solution:
Both equations are in the form $ax + by = k$. Number the equations ① and ②.

1. Eliminate y and find x.

$$2x + 3y = -2 \quad ①$$
$$3x + 7y = -6 \quad ②$$
$$\overline{14x + 21y = -14} \quad ① \times 7$$
$$-9x - 21y = 18 \quad ② \times -3$$
$$\overline{5x = 4} \quad \text{(add)}$$
$$x = \tfrac{4}{5}$$

2. Eliminate x and find y.

$$2x + 3y = -2 \quad ①$$
$$3x + 7y = -6 \quad ②$$
$$\overline{6x + 9y = -6} \quad ① \times 3$$
$$-6x - 14y = 12 \quad ② \times -2$$
$$\overline{-5y = 6} \quad \text{(add)}$$
$$5y = -6$$
$$y = -\tfrac{6}{5}$$

∴ the solution is $x = \tfrac{4}{5}$ and $y = -\tfrac{6}{5}$.

Note: This method can also be used if the solution does not contain fractions.

Exercise 1.4

Solve for x and y in questions 1–27.

1. $3x + 2y = 8$
   $2x - y = 3$

2. $5x - 3y = 14$
   $2x + y = 10$

3. $2x + y = 13$
   $x + 2y = 11$

4. $x + y = 7$
   $2x + y = 12$

5. $2x - 3y = 5$
   $x + y = -5$

6. $2x + y = 7$
   $3x - 2y = 0$

7. $x + y = 10$
   $x - y = 4$

8. $2x - 5y = 11$
   $3x + 2y = 7$

9. $2x + 3y = 12$
   $x + y = 5$

10. $2x - y = -3$
    $x - 2y = -3$

11. $2x - 3y - 14 = 0$
    $3x + 4y + 13 = 0$

12. $x - 4y - 3 = 0$
    $3x - y + 2 = 0$

13. $x - 2y = 0$
    $2(x + 3) = 3y + 5$

14. $5x + y = 19$
    $2x - y = 2(y - x)$

15. $3(x + y) + 2(y - x) = 4$
    $2(x - 2) = 3(y - 3)$

16. $3x + y = 9$
    $\dfrac{x}{2} - y = -2$

17. $2x - 5y = 19$
    $\dfrac{3x}{2} + \dfrac{4y}{3} = -1$

18. $3x - 4y = -3$
    $\dfrac{x}{2} + \dfrac{y}{3} = \dfrac{5}{2}$

19. $2x + 7y = 3$
    $x + y = \dfrac{x - 2y + 1}{2}$

20. $2(x - 5) = 3y$
    $\dfrac{2x + 1}{5} + \dfrac{x + y}{2} = 1$

21. $3x - 2y = y - 6x$
    $\dfrac{5x - 3y + 2}{2} = \dfrac{x - 2y + 4}{3}$

In questions 22–27, the solutions contain fractions.

22. $7x - 3y = 6$
    $3x - 6y = 1$

23. $5x + y = 10$
    $3x - y = 2$

24. $4x - 3y = 6$
    $2x + 6y = 13$

25. $2x + y = 2$
    $5x + 10y = 11$

26. $2x + 3y = 8$
    $2x - 3y = 2$

27. $x - y = 1$
    $3x + 5y = 7$

28. Seven books and three magazines cost €82. Two books and one magazine cost €24. Let €x be the price of a book and €y be the price of a magazine.

    (i) Write down an equation in x and y to show the price of
       (a) seven books and three magazines (b) two books and one magazine.

    (ii) Solve your two equations simultaneously.

    (iii) What is the price of (a) a book (b) a magazine?

    (iv) Calculate the price of 10 books and six magazines.

29. Here are four equations:
    A: $2x - y = 8$ B: $3x + y = 20$ C: $4x + 3y = 26$ D: $3x + 2y = 11$
    Here are four sets of (x, y) values: $(1, 4)$, $(5, 2)$, $(3, -2)$, $(-1, 23)$.
    Match each pair of (x, y) values to one of the equations, A, B, C or D.

30. The opposite sides in a parallelogram are equal in length. Use this information to calculate the values of x and y for the following parallelograms (all dimensions are in cm).

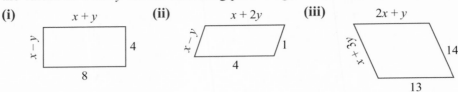

31. Solve for x and y.

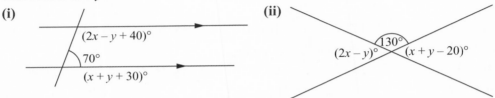

32. Andrew played a computer game. He scored 49 points by destroying five satellites and three planets and he scored 31 points by destroying three satellites and two planets. By letting x = the number of points for destroying a satellite and y = the number of points for destroying a planet, form two equations in x and y. By solving these equations, calculate the number of points Andrew scored for destroying a satellite and for destroying a planet.

33. Angela used her mobile phone to send eight text messages and four picture messages and was charged €1·44. When she sent three text messages and two picture messages she was charged €0·62. By letting x = the cost of sending a text message and y = the cost of sending a picture message, form two equations in x and y. By solving these equations, find the cost of each type of message.

34. The number in each square is the sum of the numbers in the two circles on either side of the square.

(i) P, Q, R and S are positive whole numbers. Calculate the value of P, Q, R and S.

(ii) A, B, C, D and E are positive whole numbers. If A = 3, calculate the value of B, C, D and E.

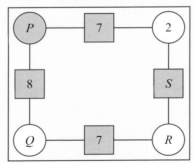
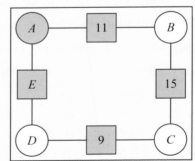

ALGEBRA

35. Each fruit symbol stands for a missing number. Calculate the value of each fruit.

(i)  = 30

(ii)  = 22

(iii)  = 40

(iv)  = 2

(v)  = 50

(vi) = 54

Factors required to solve quadratic equations

There are three types of quadratic expression we have to factorise to solve quadratic equations.

	Type	Example	Factors
1.	Quadratic trinomials	$2x^2 - 7x + 3$	$(2x - 1)(x - 3)$
2.	Taking out the HCF	$x^2 - x$	$x(x - 1)$
3.	Difference of two squares	$x^2 - 25$	$(x - 5)(x + 5)$

1. Quadratic trinomials

Factorising quadratic trinomials

Quadratic trinomials can be broken up into **two** types.

1. **Final term positive**

   When the final term is positive, the signs inside the middle of the brackets will be the **same**, either two pluses or two minuses. Keep the sign of the middle term given in the question.

   > Middle term plus: (number x + number)(number x + number)   (two pluses)
   > Middle term minus: (number x − number)(number x − number)   (two minuses)

2. **Final term negative**

   When the final term is negative, the signs inside the middle of the brackets will be **different**.

   > (number x + number)(number x − number)   (different signs)
   > or
   > (number x − number)(number x + number)   (different signs)

In both cases the factors can be found by trial and improvement. The test is to multiply the inside terms, multiply the outside terms and add the results to see if you get the middle term of the original quadratic trinomial.

ALGEBRA

EXAMPLE

Factorise each of the following.

(i) $x^2 + x - 20$ (ii) $2x^2 - 11x + 5$ (iii) $5x^2 - x - 6$

Solution:

(i) $x^2 + x - 20$

Final term $-$,

∴ the factors are $(x + \text{number})(x - \text{number})$

or

$(x - \text{number})(x + \text{number})$.

Factors of 20
1×20
2×10
4×5

Note: It is good practice to begin the trial and improvement with $(x + \text{number})(x - \text{number})$.

$(x + 2)(x - 10)$ $2x - 10x = -8x$ (no)

$(x + 4)(x - 5)$ $4x - 5x = -x$ (no, wrong sign)

On our second trial we have the correct number in front of x but of the wrong sign. So we just swap the signs:

$(x - 4)(x + 5)$ $-4x + 5x = x$ (yes, this is the middle term)

∴ $x^2 + x - 20 = (x - 4)(x + 5)$

(ii) $2x^2 - 11x + 5$

The factors of $2x^2$ are $2x$ and x.

Final term $+$ and middle term $-$,

∴ the factors are $(2x - \text{number})(x - \text{number})$.

Factors of 5
1×5

1. $(2x - 5)(x - 1)$ middle term $= -5x - 2x = -7x$ (no)

2. $(2x - 1)(x - 5)$ middle term $= -x - 10x = -11x$ (yes)

∴ $2x^2 - 11x + 5 = (2x - 1)(x - 5)$

(iii) $5x^2 - x - 6$

The factors of $5x^2$ are $5x$ and x.

Final term $-$,

∴ the factors are $(5x + \text{number})(x - \text{number})$.

Factors of 6
1×6
2×3

Note: The signs inside these brackets could be swapped.

1. $(5x + 1)(x - 6)$ middle term $= x - 30x = -29x$ (no)

2. $(5x + 6)(x - 1)$ middle term $= 6x - 5x = x$ (no, wrong sign)

The second attempt has the wrong sign of the coefficient of the middle term. Therefore, all that is needed is to swap the signs in the middle of the brackets.

3. $(5x - 6)(x + 1)$ middle term $= -6x + 5x = -x$ (yes)

$$\therefore 5x^2 - x - 6 = (5x - 6)(x + 1)$$

2. Taking out the highest common factor (HCF)

EXAMPLE

Factorise: **(i)** $x^2 - 3x$ **(ii)** $2x^2 + x$

Solution:

(i) $x^2 - 3x$
   $= x(x - 3)$ (take out the highest common factor x)

(ii) $2x^2 + x$
   $= x(2x + 1)$ (take out the highest common factor x)

3. Difference of two squares

We factorise the difference of two squares with the following steps.

1. Write each term as a perfect square with brackets.
2. Use the rule $a^2 - b^2 = (a - b)(a + b)$.
   In words: $(\text{first})^2 - (\text{second})^2 = (\text{first} - \text{second})(\text{first} + \text{second})$.

EXAMPLE

Factorise: **(i)** $x^2 - 16$ **(ii)** $x^2 - 1$

Solution:

(i) $x^2 - 16$
   $= (x)^2 - (4)^2$ (write each term as a perfect square in brackets)
   $= (x - 4)(x + 4)$ (apply the rule, (first − second)(first + second))

(ii) $x^2 - 1$
$= (x)^2 - (1)^2$ (write each term as a perfect square in brackets)
$= (x - 1)(x + 1)$ (apply the rule, (first − second)(first + second))

Exercise 1.5

Factorise each of the following quadratic trinomials in questions 1–24.

1. $x^2 + 4x + 3$
2. $x^2 - 6x + 5$
3. $x^2 + 2x - 8$
4. $x^2 - 3x - 10$
5. $x^2 - 5x + 4$
6. $x^2 + 9x + 20$
7. $x^2 + x - 12$
8. $x^2 - 2x - 15$
9. $x^2 + 13x + 30$
10. $x^2 - 3x - 28$
11. $2x^2 + 5x + 3$
12. $2x^2 - 7x + 6$
13. $2x^2 + 9x - 5$
14. $2x^2 - x - 6$
15. $3x^2 + 16x + 5$
16. $3x^2 - 22x + 7$
17. $3x^2 + 10x - 8$
18. $3x^2 - x - 2$
19. $5x^2 + 8x + 3$
20. $5x^2 - 17x + 6$
21. $5x^2 + 9x - 2$
22. $5x^2 - 19x - 4$
23. $7x^2 - 36x + 5$
24. $11x^2 + 31x - 6$

Factorise each of the following by taking out the highest common factor in questions 25–33.

25. $x^2 + 2x$
26. $x^2 - 3x$
27. $x^2 + 4x$
28. $x^2 - 5x$
29. $x^2 + 6x$
30. $x^2 - x$
31. $x^2 + x$
32. $2x^2 + 3x$
33. $2x^2 - 5x$

Factorise each of the following using the difference of two squares in questions 34–39.

34. $x^2 - 4$
35. $x^2 - 16$
36. $x^2 - 25$
37. $x^2 - 64$
38. $x^2 - 100$
39. $x^2 - 49$

40. Expand (remove the brackets) each of the following and factorise your answer.
    (i) $x^2 - 2 + 2(3x + 5)$ (ii) $3(x^2 + 4x) - 2(x^2 + 3x) - x$ (iii) $5(x^2 - 1) - 4(x^2 + 1)$

41. Write down an expression for the missing lengths, l, of each of the following rectangles.

(i) l, Area = $x^2 + 7x + 12$, $x + 3$

(ii) l, Area = $2x^2 + 11x + 5$, $x + 5$

(iii)

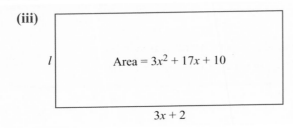

EXAMPLE

Factorise $8x^2 + 2x - 15$.

Solution:
$8x^2 + 2x - 15$

The factors of $8x^2$ are $8x$ and x or $4x$ and $2x$.

∴ the factors are:

$(8x + \text{number})(x - \text{number})$ or $(4x + \text{number})(2x - \text{number})$

Factors of 15
11 × 5
3 × 5

Note: The signs inside these brackets could be swapped.

1. $(8x + 3)(x - 5)$ middle term $= 3x - 40x = -37x$ (no)
2. $(8x + 5)(x - 3)$ middle term $= 5x - 24x = -19x$ (no)
3. $(4x + 3)(2x - 5)$ middle term $= 6x - 20x = -14x$ (no)
4. $(4x + 5)(2x - 3)$ middle term $= 10x - 12x = -2x$ (no, wrong sign)

The fourth attempt has the wrong sign of the coefficient of the middle term. Therefore, all that is needed is to swap the signs in the middle of the brackets.

5. $(4x - 5)(2x + 3)$ middle term $= -10x + 12x = 2x$ (yes)

$$\therefore 8x^2 + 2x - 15 = (4x - 5)(2x + 3)$$

Exercise 1.6

Factorise each of the following.

1. $4x^2 + 8x + 3$
2. $4x^2 - 21x + 5$
3. $6x^2 + 11x + 3$
4. $6x^2 - 13x + 2$
5. $6x^2 - x - 2$
6. $8x^2 - 25x + 3$
7. $8x^2 + 6x - 5$
8. $9x^2 - 18x + 5$
9. $10x^2 - x - 3$

Quadratic equations

> A quadratic equation is an equation in the form
> $$ax^2 + bx + c = 0$$
> where a, b and c are constants and $a \neq 0$.

Solving a quadratic equation means finding the two values of the variable which satisfy the equation. These values are called the **roots** of the equation. Sometimes the two roots are the same.

There are three types of quadratic equation we will meet on our course:

> 1. $x^2 - 2x - 3 = 0$ (three terms)
> 2. $x^2 - 5x = 0$ (no constant term)
> 3. $x^2 - 16 = 0$ (no x term)

Quadratic equations are solved with the following steps.

Method 1

> 1. Write the equation in the form $ax^2 + bx + c = 0$, where a, b and c are whole numbers.
>   (If necessary, multiply both sides by -1 to make the coefficient of x^2 positive.)
> 2. Factorise the left-hand side.
> 3. Let each factor = 0.
> 4. Solve each simple equation.

Method 2

> The roots of the quadratic equation $ax^2 + bx + c = 0$ are given by the formula:
> $$x = \frac{-b \pm \sqrt{b^2 - 4ac}}{2a}$$
>
> **Notes:** 1. The whole of the top of the right-hand side, including $-b$, is divided by $2a$.
> 2. It is often called the $-b$ or quadratic formula.
> 3. Before using the formula, make sure every term is on the left-hand side, i.e. write the equation in the form $ax^2 + bx + c = 0$.

Note: If $\sqrt{b^2 - 4ac}$ is a whole number, then $ax^2 + bx + c$ can be factorised.
The formula can still be used even if $ax^2 + bx + c$ can be factorised.

Quadratic equation type 1

 EXAMPLE

Solve for x: $3x^2 - 5x - 12 = 0$.

Solution:
Method 1: Using factors

$3x^2 - 5x - 12 = 0$

$(3x + 4)(x - 3) = 0$ \hspace{1em} (factorise the left-hand side)

$3x + 4 = 0$ \hspace{1em} or \hspace{1em} $x - 3 = 0$ \hspace{1em} (let each factor = 0)

$3x = -4$ \hspace{1em} or \hspace{1em} $x = 3$

$x = -\frac{4}{3}$ \hspace{1em} or \hspace{1em} $x = 3$ \hspace{1em} (solve each simple equation)

Method 2: Using the formula $x = \dfrac{-b \pm \sqrt{b^2 - 4ac}}{2a}$

$3x^2 - 5x - 12 = 0$

$x = \dfrac{-b \pm \sqrt{b^2 - 4ac}}{2a}$

$x = \dfrac{5 \pm \sqrt{(-5)^2 - 4(3)(-12)}}{2(3)}$ \hspace{2em} ($a = 3, b = -5, c = -12$)

$x = \dfrac{5 \pm \sqrt{25 + 144}}{6}$

$x = \dfrac{5 \pm \sqrt{169}}{6}$

$x = \dfrac{5 \pm 13}{6}$

$x = \dfrac{5 + 13}{6}$ \hspace{1em} or \hspace{1em} $x = \dfrac{5 - 13}{6}$

$x = \dfrac{18}{6}$ \hspace{1em} or \hspace{1em} $x = -\dfrac{8}{6}$

$x = 3$ \hspace{1em} or \hspace{1em} $x = -\dfrac{4}{3}$

ALGEBRA

Quadratic equation type 2

EXAMPLE

Solve for x: $x^2 + 5x = 0$.

Solution:

$$x^2 + 5x = 0 \quad \text{(every term is on the left-hand side)}$$
$$x(x + 5) = 0 \quad \text{(factorise the left-hand side)}$$
$$x = 0 \quad \text{or} \quad x + 5 = 0 \quad \text{(let each factor} = 0\text{)}$$
$$x = 0 \quad \text{or} \quad x = -5 \quad \text{(solve each simple equation)}$$

Note: It is important **not** to divide both sides by x, otherwise the root $x = 0$ is lost.

Quadratic equation type 3

EXAMPLE

Solve for x: $x^2 - 4 = 0$.

Solution:

We will use two methods to solve this quadratic equation.

Method 1

$$x^2 - 4 = 0 \quad \text{(every term is on the left-hand side)}$$
$$(x)^2 - (2)^2 = 0 \quad \text{(difference of two squares)}$$
$$(x - 2)(x + 2) = 0 \quad \text{(factorise the left-hand side)}$$
$$x - 2 = 0 \quad \text{or} \quad x + 2 = 0 \quad \text{(let each factor} = 0\text{)}$$
$$x = 2 \quad \text{or} \quad x = -2 \quad \text{(solve each simple equation)}$$

Method 2

$$x^2 - 4 = 0$$
$$x^2 = 4 \quad \text{(add 4 to both sides)}$$
$$x = \pm\sqrt{4} \quad \text{(take the square root of both sides)}$$
$$x = \pm 2$$
$$x = 2 \quad \text{or} \quad x = -2$$

Note: The examples in type 2 and type 3 could have been solved using the formula.

Exercise 1.7

Solve for x in each of the following quadratic equations in questions 1–33.

1. $(x - 2)(x - 3) = 0$
2. $(x + 5)(x - 4) = 0$
3. $(x - 3)(x + 7) = 0$
4. $x(x + 3) = 0$
5. $x(x - 5) = 0$
6. $x(x - 8) = 0$
7. $(x - 6)(x + 6) = 0$
8. $(x - 4)(x + 4) = 0$
9. $(x - 10)(x + 10) = 0$

In questions 10, 17, 21 and 23, verify your answers.

10. $x^2 - 7x + 12 = 0$
11. $x^2 + 6x + 8 = 0$
12. $x^2 - 2x - 15 = 0$
13. $x^2 - 6x + 5 = 0$
14. $x^2 + 3x - 10 = 0$
15. $x^2 + x - 20 = 0$
16. $x^2 - 6x - 7 = 0$
17. $x^2 - 9x + 14 = 0$
18. $x^2 - 5x - 24 = 0$
19. $x^2 - 4x = 0$
20. $x^2 + 6x = 0$
21. $x^2 - 2x = 0$
22. $x^2 - 9 = 0$
23. $x^2 - 25 = 0$
24. $x^2 - 1 = 0$
25. $2x^2 + 5x + 3 = 0$
26. $2x^2 - 7x + 6 = 0$
27. $2x^2 + 7x - 4 = 0$
28. $3x^2 + 10x - 8 = 0$
29. $3x^2 + 2x - 5 = 0$
30. $3x^2 - 7x + 2 = 0$
31. $5x^2 + 9x - 2 = 0$
32. $7x^2 + 5x - 2 = 0$
33. $2x^2 + 7x - 15 = 0$

In questions 34–39, first express each equation in the form $ax^2 + bx + c = 0$, where a, b and c are positive or negative whole numbers.

34. $\frac{1}{3}x^2 - x - 6 = 0$
35. $\frac{x^2}{4} - \frac{x}{2} - 2 = 0$
36. $\frac{x^2}{10} + \frac{x}{5} = \frac{3}{2}$
37. $\frac{1}{2}x^2 + x = 0$
38. $\frac{1}{4}x^2 - x = 0$
39. $\frac{1}{3}x^2 - 12 = 0$
40. Simplify $(2x - 1)(x + 1) - 2(x + 7)$ and factorise the simplified expression. Hence, solve $(2x - 1)(x + 1) - 2(x + 7) = 0$. Verify your answers.

Solve each of the following equations in questions 41–43.

41. $x(2x + 7) + 6 = 0$
42. $(x + 3)(x + 5) = 3 + x$
43. $(x - 1)^2 - 4 = 0$
44. Two whole numbers, x and $(x + 8)$, are multiplied together. The result is 84.

    (i) Write down an equation in x.
    (ii) Show that this equation can be expressed as $x^2 + 8x - 84 = 0$.
    (iii) Solve the equation to find the values of the two whole numbers.

45. A rectangular flowerbed measures $(2x + 5)$ m by $(x + 3)$ m. It has an area of 45 m². Find the value of x. Verify your answer.

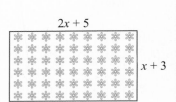

46. The triangle and the rectangle have equal areas, where $x > 0$.
   (i) Find the value of x and verify your answer.
   (ii) Find the perimeter of the triangle.

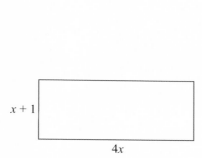
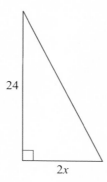

47. This rectangle is made up of four parts with areas of x^2, $5x$, $4x$ and 20 square units. If the area of the rectangle is 56 cm², calculate the value of x and verify your answer.

48. A square has length x cm. An open box is to be made by cutting 2 cm squares from each corner and folding up the sides. The volume of the box is 72 cm³. Find the dimensions of the box.

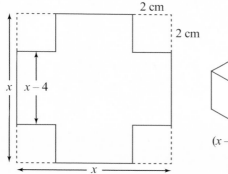
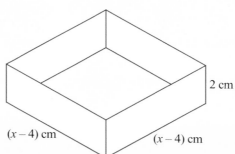

49. A closed rectangular box has a square base of side x cm. The height of the box is 2 cm. The total surface area of the box is 90 cm². Write down an equation in x to represent this information and use it to calculate x. Verify your answer.

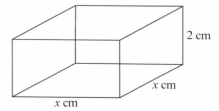

50. AT is a tangent to a circle of centre O, where T is a point on the circle.
   $|OT| = x$ cm, $|AT| = (x + 2)$ cm and $|OA| = (x + 4)$ cm.
   (i) Write down $|\angle OTA|$, giving a reason for your answer.
   (ii) If the area of $\triangle OTA = 24$ cm^2, calculate the value of x and verify your answer.
   (iii) Use another method to calculate the value of x.

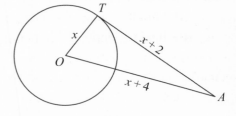

Simplifying surds

Numbers such as $\sqrt{2}, \sqrt{3}, \sqrt{5}$ and $\sqrt{7}$ are called **surds**. They cannot be written as fractions. They are also called **irrational numbers**. Below is a property of surds that we use when solving quadratic equations with the quadratic formula.

Property	Example
1. $\sqrt{ab} = \sqrt{a}\sqrt{b}$	$\sqrt{20} = \sqrt{4 \times 5} = \sqrt{4}\sqrt{5} = 2\sqrt{5}$

When simplifying surds, the key idea is to find the largest possible square number bigger than 1 that will divide evenly into the number under the square root symbol.

The square numbers greater than 1 are 4, 9, 16, 25, 36, 49, 64, 81, 100, 121, 144, etc.

You can use your calculator to help you find the largest possible square number that will divide exactly into the number under the square root symbol. Try 4, then try 9, then try 16 and so on until you find the largest possible square number that will divide exactly.

● EXAMPLE

Express (i) $\sqrt{18}$ (ii) $\sqrt{75}$ (iii) $\sqrt{80}$ in the form $a\sqrt{b}$, where $a \neq 1$.

Solution:
(i) $\sqrt{18} = \sqrt{9 \times 2} = \sqrt{9}\sqrt{2} = 3\sqrt{2}$
(ii) $\sqrt{75} = \sqrt{25 \times 3} = \sqrt{25}\sqrt{3} = 5\sqrt{3}$
(iii) $\sqrt{80} = \sqrt{16 \times 5} = \sqrt{16}\sqrt{5} = 4\sqrt{5}$

Quadratic formula

In many quadratic equations, $ax^2 + bx + c$ cannot be resolved into factors. When this happens the formula **must** be used. To save time trying to look for factors, a clue that you must use the formula is often given in the question. When the question requires an approximate answer, e.g. 'correct to two decimal places', 'correct to three significant figures', 'correct to the nearest integer' or 'express your answer in surd form', then the formula must be used.

> The roots of the quadratic equation $ax^2 + bx + c = 0$ are given by the formula
> $$x = \frac{-b \pm \sqrt{b^2 - 4ac}}{2a}.$$
>
> Notes: 1. The whole of the top of the right-hand side, including $-b$, is divided by $2a$.
> 2. It is often called the $-b$ or quadratic formula.
> 3. Before using the formula, make sure every term is on the left-hand side, i.e. write the equation in the form $ax^2 + bx + c = 0$.

EXAMPLE

Solve the equation $x^2 - 4x + 1 = 0$.

Write your answers:
(i) In the form $a \pm \sqrt{b}$, where $a, b \in \mathbb{N}$
(ii) Correct to one decimal place

Solution:
The clues, write your answers (i) in the form $a \pm \sqrt{b}$ (ii) correct to one decimal place, mean we have to use the formula.

(i) Answers in the form $a \pm \sqrt{b}$
$$x^2 - 4x + 1 = 0$$
$$x = \frac{-b \pm \sqrt{b^2 - 4a}}{2a}$$
$$x = \frac{4 \pm \sqrt{(-4)^2 - 4(1)(1)}}{2(1)} \quad (a = 1, b = -4, c = 1)$$
$$x = \frac{4 \pm \sqrt{16 - 4}}{2}$$
$$x = \frac{4 \pm \sqrt{12}}{2}$$

$$x = \frac{4 \pm 2\sqrt{3}}{2} \qquad (\sqrt{12} = \sqrt{4 \times 3} = \sqrt{4}\sqrt{3} = 2\sqrt{3})$$

$$x = \frac{4}{2} \pm \frac{2\sqrt{3}}{2}$$

$$x = 2 \pm \sqrt{3}$$

(ii) Correct to one decimal place

$x = 2 \pm \sqrt{3}$

$x = 2 + \sqrt{3}$ or $x = 2 - \sqrt{3}$

$x = 2 + 1\cdot732050808$ or $x = 2 - 1\cdot732050808$

$x = 3\cdot732050808$ or $x = 0\cdot2679491924$

$x = 3\cdot7$ or $x = 0\cdot3$, correct to one decimal place

Exercise 1.8

Write questions 1–10 in the form $a\sqrt{b}$, where $a \neq 1$.

1. $\sqrt{8}$ 2. $\sqrt{24}$ 3. $\sqrt{45}$ 4. $\sqrt{32}$ 5. $\sqrt{27}$
6. $\sqrt{48}$ 7. $\sqrt{54}$ 8. $\sqrt{125}$ 9. $\sqrt{90}$ 10. $\sqrt{50}$

Solve each of the following equations in questions 11–19, writing your answers (i) in the form $a \pm \sqrt{b}$ (ii) correct to one decimal place.

11. $x^2 - 2x - 4 = 0$ 12. $x^2 + 2x - 2 = 0$ 13. $x^2 - 4x - 1 = 0$
14. $x^2 + 6x + 7 = 0$ 15. $x^2 - 6x + 4 = 0$ 16. $x^2 + 8x + 13 = 0$
17. $x^2 + 10x + 23 = 0$ 18. $x^2 - 10x + 18 = 0$ 19. $x^2 + 12x + 33 = 0$

In questions 20–22, write your answer in the form $\dfrac{a \pm \sqrt{b}}{c}$.

20. $2x^2 - 2x - 1 = 0$ 21. $4x^2 + 2x - 1 = 0$ 22. $9x^2 + 6x - 1 = 0$

Solve each of the following equations in questions 23–25, giving your answers correct to two decimal places.

23. $x^2 - 4x - 14 = 0$ 24. $2x^2 - x - 2 = 0$ 25. $5x^2 + 7x - 4 = 0$
26. (i) A rectangle has dimensions $(x + 3)$ m by $(x + 1)$ m. If the area of the rectangle is 10 m², find the value of x correct to two decimal places.

(ii) Calculate the error by using your value of x.

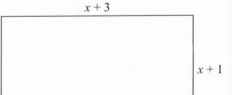

27. (i) Solve for x, $x(x - 2) = 3 + 2x$ and give your solution in the from $a \pm b$, where $a, b \in \mathbb{N}$.
(ii) Write one of your solutions correct to two decimal places. (iii) Using this value, show that the difference between the values of the left-hand side and the right-hand side of the given equation is less than 0.1.

28. The diagram shows a trapezium. The measurements on the diagram are in cm.
The lengths of the parallel sides are x cm and 14 cm.
The height of the trapezium is $2x$ cm.
The area of the trapezium is 100 cm².

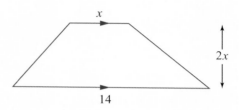

(i) Show that $x^2 + 14x - 100 = 0$.
(ii) Find the value of x. Give your answer correct to one decimal place.
(iii) Calculate the error by using this value of x.

Constructing a quadratic equation when given its roots

This is the reverse process to solving a quadratic equation by using factors.

> **EXAMPLE**
>
> Find a quadratic equation with roots (i) -2 and 3 (ii) $\frac{2}{3}$ and $-\frac{1}{5}$.
> Write your answers in the form $ax^2 + bx + c = 0$, $a, b, c \in \mathbb{Z}$.
>
> **Solution:**
>
> (i) Roots -2 and 3
> Let $x = -2$ and $x = 3$
> $x + 2 = 0$ and $x - 3 = 0$
> $(x + 2)(x - 3) = 0$
> $x^2 - 3x + 2x - 6 = 0$
> $x^2 - x - 6 = 0$
>
> (ii) Roots $\frac{2}{3}$ and $-\frac{1}{5}$
> Let $x = \frac{2}{3}$ and $x = -\frac{1}{5}$
> $3x = 2$ and $5x = -1$
> $3x - 2 = 0$ and $5x + 1 = 0$
> $(3x - 2)(5x + 1) = 0$
> $15x^2 + 3x - 10x - 2 = 0$
> $15x^2 - 7x - 2 = 0$

Note: $0 \times 0 = 0$

Exercise 1.9

In questions 1–20, construct a quadratic equation with roots.
(In each case, write your answer in the form $ax^2 + bx + c = 0$, $a, b, c \in \mathbb{Z}$.)

1. 2, 3
2. −1, 2
3. −2, 5
4. −1, 4
5. −3, −2
6. 4, 5
7. −3, 4
8. −8, 3

9. $-3, 3$	10. $2, 2$	11. $-2, 0$	12. $0, 5$
13. $-1, 1$	14. $\frac{1}{2}, 3$	15. $-\frac{1}{3}, 2$	16. $-3, \frac{1}{2}$
17. $-1, \frac{5}{2}$	18. $\frac{1}{3}, \frac{1}{2}$	19. $\frac{1}{3}, -\frac{2}{3}$	20. $\frac{1}{2}, \frac{3}{4}$

21. The equation $x^2 + mx + n = 0$ has roots -3 and 5. Find the values of m and n.

22. The equation $ax^2 + bx + c = 0$ has roots $-\frac{1}{2}$ and $\frac{2}{5}$. Find one set of values of a, b and c, where $a, b, c \in \mathbb{Z}$.

Simultaneous equations, one linear and one quadratic

The solution of a pair of simultaneous equations where one is linear (line) and one is quadratic (curve) represents the points of intersection of a line and a curve. Graphing a line and a curve will lead to three possibilities.

1.

2.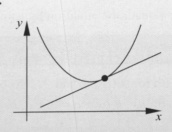

3.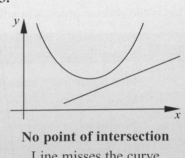

Two points of intersection
Line meets the curve in two different points

One point of intersection
Line is a tangent to the curve

No point of intersection
Line misses the curve

The **method of substitution** is used to solve between a linear equation and a quadratic equation. The method involves three steps.

1. From the linear equation, express one variable in terms of the other.
2. Substitute this into the quadratic equation and solve.
3. Substitute **separately** the value(s) obtained in step 2 into the linear equation in step 1 to find the corresponding value(s) of the other variable.

EXAMPLE 1

Solve for x and y: $x + 3 = 2y$ and $xy - 7y + 8 = 0$.

Solution:

$$x + 3 = 2y \quad \text{and} \quad xy - 7y + 8 = 0$$

1. $x + 3 = 2y$ (get x on its own from the linear equation)

   $x = 2y - 3$ (x on its own)

2. $xy - 7y + 8 = 0$

   $(2y - 3)y - 7y + 8 = 0$ (put in $(2y - 3)$ for x)

   $2y^2 - 3y - 7y + 8 = 0$ (remove the brackets)

   $2y^2 - 10y + 8 = 0$ (simplify the left-hand side)

   $y^2 - 5y + 4 = 0$ (divide both sides by 2)

   $(y - 1)(y - 4) = 0$ (factorise the left-hand side)

   $y - 1 = 0$ or $y - 4 = 0$ (let each factor $= 0$)

   $y = 1$ or $y = 4$ (solve each simple equation)

3. Substitute $y = 1$ and $y = 4$ separately into the linear equation.

   $x = 2y - 3$

$y = 1$	$y = 4$
$x = 2y - 3$	$x = 2y - 3$
$x = 2(1) - 3$	$x = 2(4) - 3$
$x = 2 - 3$	$x = 8 - 3$
$x = -1$	$x = 5$
$x = -1, y = 1$	$x = 5, y = 4$

   $\therefore$ the solutions are $x = -1$ and $y = 1$ or $x = 5$ and $y = 4$.

   The line and the curve meet at the points $(-1, 1)$ and $(5, 4)$.

EXAMPLE 2

Solve for x and y: $x + y = 7$ and $x^2 + y^2 = 29$.

Solution:

$x + y = 7$ and $x^2 + y^2 = 29$

1. $x + y = 7$ (get x or y on its own from the linear equation)

   $y = 7 - x$ (y on its own)

2. $\quad x^2 + y^2 = 29$

   $x^2 + (7 - x)^2 = 29$ (put in $(7 - x)$ for y)

   $x^2 + 49 - 14x + x^2 = 29$ $\quad ((7 - x)^2 = 49 - 14x + x^2)$

   $2x^2 - 14x + 49 = 29$ (simplify the left-hand side)

   $2x^2 - 14x + 20 = 0$ (subtract 29 from both sides)

   $x^2 - 7x + 10 = 0$ (divide both sides by 2)

   $(x - 2)(x - 5) = 0$ (factorise the left-hand side)

   $x - 2 = 0$ or $x - 5 = 0$ (let each factor $= 0$)

   $x = 2$ or $x = 5$ (solve each simple equation)

3. Substitute $x = 2$ and $x = 5$ separately into the linear equation.

$y = 7 - x$	
$x = 2$	$x = 5$
$y = 7 - x$	$y = 7 - x$
$y = 7 - 2$	$y = 7 - 5$
$y = 5$	$y = 2$
$x = 2, y = 5$	$x = 5, y = 2$

∴ the solutions are

$x = 2$ and $y = 5$ or $x = 5$ and $y = 2$.

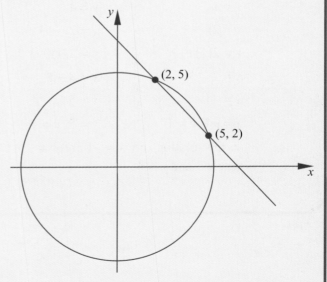

The line and curve meet at the points (2, 5) and (5, 2).

Exercise 1.10

1. Solve each of the following pairs of simultaneous equations.
   (i) $y = x - 5$ and $y = x^2 - 5x + 3$
   (ii) $y = 3x + 2$ and $y = x^2 + x - 1$

2. In the following, $f(x)$ represents the line and $g(x)$ represents the curve. In each case, find the coordinates of the points of intersection.
   (i) $f(x) = x + 3 \quad g(x) = x^2 - 2x - 7$

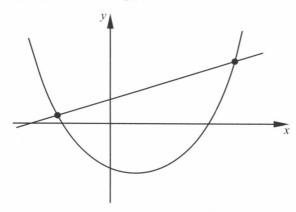

   (ii) $f(x) = 2x - 3 \quad g(x) = 9 + x - x^2$

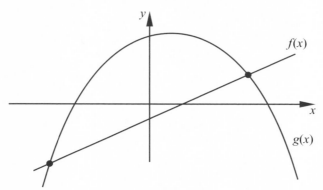

3. Verify that the line $4x - y - 7 = 0$ is a tangent to the curve $y = x^2 - 2x + 2$ and find the coordinates of the point of contact.

Solve each of the following pairs of simultaneous equations in questions 4–6.

4. $x + y = 5$
   $xy = 6$

5. $x - y = 1$
   $xy = 2$

6. $2x + y - 6 = 0$
   $xy = 4$

Write down the expansion of each of the following in questions 7–14.

7. $(x + 3)^2$
8. $(x - 2)^2$
9. $(y - 1)^2$
10. $(y + 3)^2$
11. $(2 + x)^2$
12. $(1 - 2x)^2$
13. $(3 - 2x)^2$
14. $(3 - 5y)^2$

Solve each of the following for x and y in questions 15–23.

15. $x + y = 3$
    $x^2 + y^2 = 17$

16. $x - y = 1$
    $x^2 + y^2 = 25$

17. $y = x + 1$
    $x^2 + y^2 = 1$

18. $x - 2y = 0$
    $x^2 + y^2 = 20$

19. $x = 2y + 5$
    $x^2 + y^2 = 25$

20. $x + y = 5$
    $x^2 + y^2 = 13$

21. $x - y - 4 = 0$
    $y^2 + 3x = 16$

22. $x = 3 - y$
    $x^2 - y^2 + 3 = 0$

23. $2x + y = 1$
    $x^2 + xy + y^2 = 7$

24. (i) Solve for x and y: $y = 10 - 2x$ and $x^2 + y^2 = 25$.

    (ii) Hence, find the two possible values of $x^3 + y^3$.

25. A rectangle has dimensions as shown. Its perimeter is 14 cm and its area is 12 cm².

    (i) Derive the equations $x + y = 7$ and $xy = 12$.

    (ii) Solve the equations and deduce the dimensions of the rectangle.

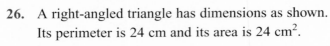

26. A right-angled triangle has dimensions as shown. Its perimeter is 24 cm and its area is 24 cm².

    (i) Derive the equations $x + y = 12$ and $xy - x = 24$, where $x < y$.

    (ii) Solve the equations and deduce the dimensions of the triangle.

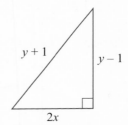

27. The graph below shows a line $f(x) = x + 1$ and a curve $g(x) = -x^2 + 10x - 7$.

    The line represents the top flat surface of a hill and the curve represents the flight path of a projectile.

    The projectile is fired from point A and it lands at point B.

    (i) Find the coordinates of points A and B.

    (ii) Calculate $|AB|$.

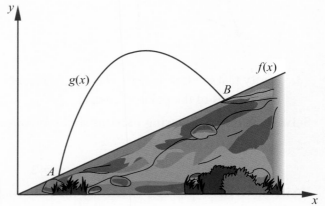

28. The diagram shows the cross-section of a road tunnel. The equation of the inner shape of the tunnel is given by $x^2 + 4y^2 = 40$. The tunnel is closed and a security barrier, $[PQ]$, is erected to prevent vehicles entering the tunnel. The equation of the security barrier is given by $x - 4y + 10 = 0$. Bolts at an angle hold the barrier at positions P and Q, as shown. Find the coordinates of P and Q.

29. A model railway bridge is supported by straight bars connected to an arch, as shown. The arch is resting on the ground. The ground is represented by the x-axis and the arch is symmetrical about the y-axis.

The equations of the straight bars are given by $2x - y + 6 = 0$ and $2x + y - 6 = 0$.
The equation of the arched frame is given by $20x^2 + y^2 = 180$.
The equation of the railway track is given by $y = 16$.
Find the coordinates of the points where:

(i) The straight bars meet the arched frame

(ii) The straight bars meet the railway line

Inequalities

The four inequality symbols are:

> 1. $>$ means greater than
> 2. $\geq$ means greater than or equal to
> 3. $<$ means less than
> 4. $\leq$ means less than or equal to

Algebraic expressions that are linked by one of the four inequality symbols are called **inequalities**. For example, $3x - 1 \leq 11$ and $-3 < 2x - 1 \leq 7$ are inequalities.

Solving inequalities is exactly the same as solving equations, with the following exception:

> Multiplying or dividing both sides of an inequality by a **negative** number **reverses** the direction of the inequality symbol.
> That is:
>
> $>$ changes to $<$ $\geq$ changes to $\leq$
>
> $<$ changes to $>$ $\leq$ changes to $\geq$

For example, $5 > -3$ is true. If we multiply both sides by -1, it gives $-5 < 3$, which is also true.

Solving an inequality means finding the values of x that make the inequality true.

The following rules apply to graphing inequalities on a number line:

> Number line for $x \in \mathbb{N}$ or $x \in \mathbb{Z}$, use **dots**.
>
> Number line for $x \in \mathbb{R}$, use a **full** heavy line.

Note: Inequalities can be turned around. For example:

$5 \leq x$ means the same as $x \geq 5$

$8 \geq x \geq 3$ means the same as $3 \leq x \leq 8$

Single variable linear inequalities

EXAMPLE 1

Find the solution set of $14 - 3x \geq 2$, $x \in \mathbb{N}$ and graph your solution on the number line.

Solution:

$$14 - 3x \geq 2$$
$$-3x \geq -12 \quad \text{(subtract 14 from both sides)}$$
$$3x \leq 12 \quad \text{(multiply both sides by } -1 \text{ and reverse the inequality symbol)}$$
$$x \leq 4 \quad \text{(divide both sides by 3)}$$

As $x \in \mathbb{N}$, this is the set of natural numbers less than or equal to 4.
Thus, the values of x are 1, 2, 3 and 4.

Number line:

Note: As $x \in \mathbb{N}$, dots are used on the number line.

EXAMPLE 2

Solve the inequality $3x - 2 < 7x + 6$, $x \in \mathbb{R}$ and illustrate the solution on the number line.

Solution:

$$3x - 2 < 7x + 6$$
$$3x < 7x + 8 \quad \text{(add 2 to both sides)}$$
$$-4x < 8 \quad \text{(subtract } 7x \text{ from both sides)}$$
$$4x > -8 \quad \text{(multiply both sides by } -1 \text{ and reverse the inequality)}$$
$$x > -2 \quad \text{(divide both sides by 4)}$$

This is the set of real numbers greater than -2 (-2 is **not** included).

Number line:

A circle is put around -2 to indicate that it is **not** part of the solution.

Note: As $x \in \mathbb{R}$, we use full heavy shading on the number line.

EXAMPLE 3

(i) Find A, the solution set of $7x - 1 \leq 27$, $x \in \mathbb{Z}$.

(ii) Find B, the solution set of $\dfrac{5 - 3x}{2} \leq 4$, $x \in \mathbb{Z}$.

(iii) Find $A \cap B$ and graph your solution on the number line.

Solution:

(i) $A: 7x - 1 \leq 27$
$7x \leq 28$
$x \leq 4$

(ii) $B: \dfrac{5 - 3x}{2} \leq 4$
$5 - 3x \leq 8$ (multiply both sides by 2)
$-3x \leq 3$
$3x \geq -3$
$x \geq -1$

(iii) $A \cap B$: combining the two inequalities:
$-1 \leq x \leq 4$

This is the set of positive and negative whole numbers between -1 and 4, including -1 and 4.

Number line:

Note: As $x \in \mathbb{Z}$, dots are used on the number line.

Double inequalities

A double inequality is one like $-3 \leq 2x + 1 \leq 7$.
There are two methods for solving double inequalities.

Method 1

> Whatever we do to one part, we do the same to all three parts.

Method 2

> 1. Write the double inequality as two separate simple inequalities.
> 2. Solve each simple inequality and combine their solutions.

EXAMPLE

Solve the inequality $-6 \leq 5x - 1 < 9$, $x \in \mathbb{R}$.
Graph your solution on a number line.

Solution:

Method 1: Do the same to all three parts.

$$-6 \leq 5x - 1 < 9$$
$$-5 \leq 5x < 10 \qquad \text{(add 1 to each part)}$$
$$-1 \leq x < 2 \qquad \text{(divide each part by 5)}$$

Method 2: Write the double inequality as two separate inequalities.

1st inequality		2nd inequality
$-6 \leq 5x - 1$	and	$5x - 1 < 9$
$-6 \leq 5x - 1$		$5x - 1 < 9$
$-5 \leq 5x$		$5x < 10$
$-1 \leq x$		$x < 2$

$$-1 \leq x < 2 \qquad \text{(combining solutions)}$$

Number line:

A circle is put around 2 to indicate that 2 is **not** included in the solution.

Note: As $x \in \mathbb{R}$, we use full heavy shading on the number line.

Exercise 1.11

Solve each of the following inequalities in questions 1–18. In each case, graph your solution on the number line.

1. $2x + 1 \geq 7$, $x \in \mathbb{R}$
2. $3x + 1 \leq 13$, $x \in \mathbb{R}$
3. $5x - 3 > 3x + 1$, $x \in \mathbb{N}$
4. $8x - 1 < 3x + 9$, $x \in \mathbb{R}$
5. $2x - 1 \geq 4x - 7$, $x \in \mathbb{R}$
6. $6x - 10 \leq 9x + 5$, $x \in \mathbb{Z}$
7. $2(x + 4) < 2 - x$, $x \in \mathbb{R}$
8. $4(x - 2) \geq 5(2x - 1) - 9$, $x \in \mathbb{R}$
9. $\dfrac{x}{2} + \dfrac{x}{3} \geq \dfrac{5}{6}$, $x \in \mathbb{R}$
10. $\dfrac{3x}{5} - \dfrac{x}{2} \leq \dfrac{3}{10}$, $x \in \mathbb{R}$

11. $2x + 1 \leq 5$, $x \in \mathbb{N}$
12. $4x - 15 \leq 1$, $x \in \mathbb{N}$
13. $13 - 2x < 3$, $x \in \mathbb{N}$
14. $12 - 5x > 2$, $x \in \mathbb{N}$
15. $4 \leq 2x \leq 10$, $x \in \mathbb{N}$
16. $-4 \leq 3x - 1 < 11$, $x \in \mathbb{Z}$
17. $-7 \leq 5x + 3 < 18$, $x \in \mathbb{R}$
18. $-5 < 4x + 7 \leq 35$, $x \in \mathbb{R}$
19. (i) Find the solution set of (a) $A: x - 1 \geq 2$, $x \in \mathbb{R}$ (b) $B: x + 4 \leq 9$, $x \in \mathbb{R}$.
    (ii) Find $A \cap B$ and graph your solution on the number line.
20. (i) Find the solution set of (a) $H: 2x - 3 \leq 5$, $x \in \mathbb{R}$ (b) $K: 3x + 2 \geq -4$, $x \in \mathbb{R}$.
    (ii) Find $H \cap K$ and graph your solution on the number line.
21. (i) Find the solution set E of $2x + 7 \leq 19$, $x \in \mathbb{R}$.
    (ii) Find the solution set H of $3 - 2x \leq 11$, $x \in \mathbb{R}$.
    (iii) Find $E \cap H$.
22. (i) Find the solution set H of $2x + 5 \geq -1$, $x \in \mathbb{R}$.
    (ii) Find the solution set K of $7 - 3x \geq 4$, $x \in \mathbb{R}$.
    (iii) Find $H \cap K$ and graph your solution on a number line.
23. (i) Find A, the solution set of $3x - 2 \leq 4$, $x \in \mathbb{Z}$.
    (ii) Find B, the solution set of $\dfrac{1 - 3x}{2} < 5$, $x \in \mathbb{Z}$.
    (iii) List the elements of $A \cap B$.
24. (i) Find the solution set E of $9 - 2x \geq 7$, $x \in \mathbb{N}$.
    (ii) Find the solution set H of $\frac{1}{4}x - \frac{1}{3} \leq \frac{5}{12}$, $x \in \mathbb{N}$.
    (iii) Write down the elements of the set $H \setminus E$.
25. Find the smallest natural number k such that $2x + 4(x + 3) + 7(2x + 4) < 20(x + k)$.
26. Show that there are no real numbers which simultaneously satisfy the two inequalities $2x - 1 \geq 9$ and $3x + 2 \leq 14$. Explain your answer.
27. Write down the values of x that satisfy each of the following.
    (i) $x - 1 \leq 4$, where x is a positive, even number.
    (ii) $x + 4 < 6$, where x is a positive, odd number.
    (iii) $2x - 13 < 37$, where x is a square number.
    (iv) $2x + 5 < 27$, where x is a prime number.
28. Aishling said, 'I thought of a whole number, multiplied it by 5 then subtracted 3. The answer was between 11 and 23.' List the whole numbers that Aishling could have used.
29. The lengths of the sides of a triangle are $(2x + 3)$ cm, $(2x + 2)$ cm and x cm. Find the range of values of x for which this triangle exists.

30. Match the words with the correct inequality shown on the right.

    (i) x is less than 6   (ii) x is greater than or equal to 6

    (iii) x is greater than 6   (iv) x is less than or equal to 6

    (v) x is at least 6   (vi) x has a maximum value of 6

    (vii) x is at most 6   (viii) x has a minimum value of 6

A: $x \geq 6$
B: $x > 6$
C: $x \leq 6$
D: $x < 6$

31. (i) A rectangle has dimensions as shown. Explain why $x > 2$.

    (ii) The number of centimetres in its perimeter is greater than the number of square centimetres in its area. Write an equality to represent this information and solve it to find the range of values of x.

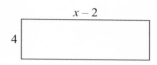

32. A family has four children. The table shows some information about their ages in years.

Name	Andrew	Bernadette	Catherine	Dermot
Age in years	n	$2n + 6$	14	22

    (i) Bernadette is older than Catherine but younger than Dermot. Calculate all Andrew's possible ages.

    (ii) Could any of these children be twins? Justify your answer.

33. The diagram shows a map of an island. A gold coin is buried at a place where the x and y coordinates are positive whole numbers. Use the clues to work out the coordinates where the gold coin is buried.

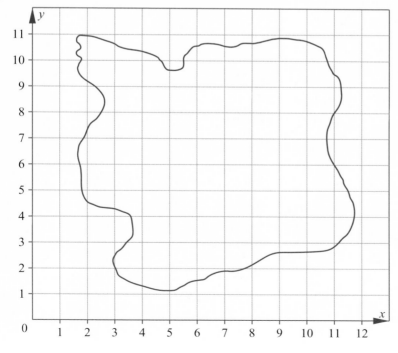

Clues:

    (i) $x > 7$

    (ii) $y > 6$

    (iii) $x + y = 17$

    (iv) One of x and y is prime and the other is not

Changing the subject of a formula

When we rearrange a formula so that one of the variables is given in terms of the others, we are **changing the subject of the formula**.
Changing the subject of a formula is solved with the following method:

> Whatever you do to one side, you must do **exactly the same** to the other side.

Note: Keep balance in mind. Whatever letter comes after the word 'express' is to be on its own.

EXAMPLE

(i) Given that $px - q = r$, express x in terms of p, q and r, where $p \neq 0$.

(ii) Given that $u^2 + 2as = v^2$, express s in terms of v, u and a.

(iii) Express x in terms of r and s when $r - \dfrac{x}{s} = 1$, $s \neq 0$.

(iv) Express t in terms of p and q when $p = \dfrac{q - t}{3t}$, $t \neq 0$.

Solution:

(i) $px - q = r$

$\quad px = r + q$ \quad (add q to both sides)

$\quad \dfrac{px}{p} = \dfrac{r + q}{p}$ \quad (divide both sides by p)

$\quad x = \dfrac{r - q}{p}$ \quad (simplify the left-hand side)

(ii) $u^2 + 2as = v^2$

$\quad 2as = v^2 - u^2$ \quad (subtract u^2 from both sides)

$\quad \dfrac{2as}{2a} = \dfrac{v^2 - u^2}{2a}$ \quad (divide both sides by $2a$)

$\quad s = \dfrac{v^2 - u^2}{2a}$ \quad (simplify the left-hand side)

(iii) $\quad r - \dfrac{x}{s} = 1$

$\quad\quad sr - \dfrac{sx}{s} = s(1)$ $\quad$ (multiply each part by s)

$\quad\quad sr - x = s$ $\quad\quad\quad\left(\dfrac{sx}{s} = x \text{ and } s(1) = s\right)$

$\quad\quad -x = s - sr$ $\quad$ (subtract sr from both sides)

$\quad\quad x = -s + sr$ $\quad$ (multiply both sides by -1)

(iv) $\quad p = \dfrac{q - t}{3t}$

$\quad\quad 3tp = \dfrac{3t(q - t)}{3t}$ $\quad$ (multiply both sides by $3t$)

$\quad\quad 3tp = q - t$ $\quad$ (simplify the right-hand side)

$\quad\quad 3tp + t = q$ $\quad$ (add t to both sides)

$\quad\quad t(3p + 1) = q$ $\quad$ (take out common factor t on the left-hand side)

$\quad\quad \dfrac{t(3p + 1)}{3p + 1} = \dfrac{q}{3p + 1}$ $\quad$ (divide both sides by $(3p + 1)$)

$\quad\quad t = \dfrac{q}{3p + 1}$ $\quad$ (simplify the left-hand side)

Exercise 1.12

Change each of the formulae in questions 1–30 to express the letter in square brackets in terms of the others.

1. $2a - b = c$ $\quad$ [a]
2. $3p + q = r$ $\quad$ [p]
3. $ab - c = d$ $\quad$ [a]
4. $u + at = v$ $\quad$ [t]
5. $3a + 2b = 5c$ $\quad$ [b]
6. $3q - 4p = 2r$ $\quad$ [q]
7. $2(a - b) = c$ $\quad$ [a]
8. $a(b - c) = d$ $\quad$ [b]
9. $x(y + z) = w$ $\quad$ [y]
10. $\tfrac{1}{2}a = b$ $\quad$ [a]
11. $\dfrac{b}{2} + c = a$ $\quad$ [b]
12. $s + \dfrac{t}{3} = r$ $\quad$ [t]
13. $\dfrac{a}{2} + \dfrac{b}{3} = c$ $\quad$ [a]
14. $\dfrac{p + q}{2} = r$ $\quad$ [q]
15. $r = \tfrac{1}{3}(p - q)$ $\quad$ [p]
16. $a = \dfrac{b - 2c}{3}$ $\quad$ [c]
17. $x + \dfrac{w}{y} = z$ $\quad$ [w]
18. $2p + \dfrac{3q}{r} = s$ $\quad$ [q]
19. $\dfrac{p - 3r}{q} = 5$ $\quad$ [p]
20. $s = \dfrac{p}{q} + \dfrac{r}{q}$ $\quad$ [q]
21. $\dfrac{2a}{b} - \dfrac{3c}{b} = d$ $\quad$ [b]

22. $\frac{1}{2}(3a + b) = \frac{1}{3}c$ [a]

23. $u^2 + 2as = v^2$ [a]

24. $a = \frac{b}{4} - 2c$ [c]

25. $\frac{1}{2}at^2 = s$ [a]

26. $v = \frac{1}{3}\pi r^2 h$ [h]

27. $s = ut + \frac{1}{2}at^2$ [a]

28. $r = \frac{1}{s} + t$ [s]

29. $p + \frac{t}{q} = r$ [q]

30. $x - \frac{y}{z} = w$ [z]

31. The formula for finding the speed, v, of a body after accelerating for t seconds is given by $v = u + at$. **(i)** Express t in terms of u, v and a. **(ii)** Find t when $v = 650$, $u = 50$ and $a = 15$.

32. **(i)** The area of a trapezium is given by
    $A = \left(\dfrac{a+b}{2}\right)h$. Express h in terms of A, a and b.
    (ii) Find the value of h when $A = 150$, $a = 10$ and $b = 15$.

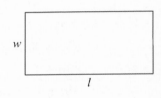

33. The diagram shows a rectangle of length l and width w. Its perimeter is P and its area is A. Express:

    (i) P in terms of l and w **(ii)** w in terms of P and l

    (iii) A in terms of l and w **(iv)** w in terms of A and l

    (v) Hence, express A in terms of P and l.

34. A farmer wants to fence off part of his garden. He buys 18 m of fencing and uses it to make three sides of a rectangle, using a fence as the fourth side, as shown. The length of one side of the rectangle is x m and the length of the other side is y m, where $x < y$.

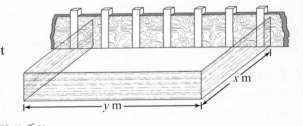

    (i) Write down an equation in x and y to represent this information.

    (ii) Express y in terms of x.

    (iii) If the area of the rectangle is 40 m², explain why $xy = 40$.

    (iv) Using your answer from part **(ii)**, write down an equation in x only to represent the area of the rectangle and solve this equation to find the values of x and y.

35. Temperatures can be measured in degrees Celsius (°C), degrees Fahrenheit (°F) or degrees Kelvin (°K). The relationships between the scales of temperature are given by

    $$C = \frac{5(F-32)}{9} \quad \text{and} \quad K = C + 273.$$

    (i) Express F in terms of **(a)** C **(b)** K.

    (ii) Hence or otherwise, calculate F when **(a)** $C = 10$ and **(b)** $K = 12$.

Notation for indices

We use a shorthand called **index notation** to indicate repeated multiplication.

For example, we write $2 \times 2 \times 2 \times 2 \times 2$ as 2^5.
This is read as '2 to the power of 5'.

The power or index simply tells you how many times a number is multiplied by itself.

> 2 is the **base**.
>
> 5 is the **index** or **power**.

Rules of indices

1. $a^m \times a^n = a^{m+n}$ Example: $2^4 \times 2^3 = 2^{4+3} = 2^7$
   Multiplying powers of the same number: **add** the indices.

2. $\dfrac{a^m}{a^n} = a^{m-n}$ Example: $\dfrac{3^9}{3^5} = 3^{9-5} = 3^4$
   Dividing powers of the same number: **subtract** the index on the bottom from the index on top.

3. $(a^m)^n = a^{mn}$ Example: $(4^5)^3 = 4^{5 \times 3} = 4^{15}$
   Raising the power of a number to a power, multiply the indices.

4. $(ab)^m = a^m b^m$ Example: $(2 \times 3)^5 = 2^5 \times 3^5$
   Raising a product to a power, every factor is raised to the power.

5. $\left(\dfrac{a}{b}\right)^m = \dfrac{a^m}{b^m}$ Example: $\left(\dfrac{2}{5}\right)^3 = \dfrac{2^3}{5^3}$
   Raising a fraction to a power, **both** top and bottom are raised to the power.

6. $a^0 = 1$ Example: $4^0 = 1$
   Any number to the power of zero is 1.

7. $a^{-m} = \dfrac{1}{a^m}$ Example: $5^{-2} = \dfrac{1}{5^2}$
   A number with a negative index is equal to its reciprocal with a positive index.

Note: If a term is brought from the top to the bottom of a fraction (or vice versa), the sign of its index is changed.

8. $a^{m/n} = (a^{1/n})^m$ Example: $32^{3/5} = (32^{1/5})^3$
   Take the root first and then raise to the power (or vice versa).

$8^{1/3}$ means the number that multiplied by itself three times will equal 8.

Thus, $8^{1/3} = 2$, as $2 \times 2 \times 2 = 8$.

Similarly, $25^{1/2} = 5$, as $5 \times 5 = 25$ and $81^{1/4} = 3$, as $3 \times 3 \times 3 \times 3 = 81$.

Note: $\sqrt{a} = a^{1/2}$, for example, $\sqrt{16} = 16^{1/2} = 4$.
Also, $\sqrt{a}\sqrt{a} = a^{1/2} \cdot a^{1/2} = a^{1/2+1/2} = a^1 = a$.

Alternative notation: $a^{1/n} = \sqrt[n]{a}$ Example: $8^{1/3} = \sqrt[3]{8}$
$a^{m/n} = \sqrt[n]{a^m}$ Example: $32^{2/5} = \sqrt[5]{32^2}$

When dealing with fractional indices, the calculations are simpler if the root is taken first and the result is raised to the power.

For example, $16^{3/4} = (16^{1/4})^3 = (2)^3 = 8$

               (root first)   (power next)

Using a calculator

A calculator can be used to evaluate an expression such as $32^{3/5}$.

$$(\boxed{} \; 32 \; \boxed{y^x} \; (3 \div 5) \; \boxed{=})$$

The calculator will give an answer of 8.

However, there are problems when dealing with negative indices or raising a fraction to a power, as the calculator can give the answer as a decimal.

For example, $8^{-2/3} = \dfrac{1}{8^{2/3}} = \dfrac{1}{(8^{1/3})^2} = \dfrac{1}{(2)^2} = \dfrac{1}{4}$

Using a calculator,

$$(\boxed{} \; 8 \; \boxed{y^x} \; \boxed{+/-} \; (2 \div 3) \; \boxed{=}) \quad \text{gives an answer of } 0.25.$$

Note: $\frac{1}{4} = 0.25$

Also, $\left(\dfrac{8}{27}\right)^{2/3} = \dfrac{8^{2/3}}{27^{2/3}} = \dfrac{(8^{1/3})^2}{(27^{1/3})^2} = \dfrac{(2)^2}{(3)^2} = \dfrac{4}{9}$

Using a calculator,

$$(\boxed{} \; (\; 8 \div 27 \;) \; \boxed{y^x} \; (2 \div 3) \; \boxed{=}) \quad \text{gives an answer of } 0.444444444 \ldots$$

Note: $\frac{4}{9} = 0.444444444\ldots$

Avoid using a calculator with negative indices or when raising a fraction to a power.

EXAMPLE 1

Write the following without indices: **(i)** 6^{-2} **(ii)** $81^{1/2}$ **(iii)** $27^{4/3}$ **(iv)** $32^{3/5}$

Solution:

(i) $6^{-2} = \dfrac{1}{6^2} = \dfrac{1}{36}$

(ii) $81^{1/2} = 9$

(iii) $27^{4/3} = (27^{1/3})^4 = (3)^4 = 81$

(iv) $32^{3/5} = (32^{1/5})^3 = (2)^3 = 8$

EXAMPLE 2

Write the following as a power of 2: (i) 8 (ii) $8^{4/3}$ (iii) $\sqrt{8}$ (iv) 4^{-3}

Solution:

(i) $8 = 2^3$

(ii) **Method 1**
$8^{4/3} = (2^3)^{4/3} = 2^{3 \times \frac{4}{3}} = 2^4$

(ii) **Method 2**
$8^{4/3} = (8^{1/3})^4 = (2)^4 = 16 = 2^4$

(iii) $\sqrt{8} = (8)^{1/2} = (2^3)^{1/2} = 2^{3 \times \frac{1}{2}} = 2^{3/2}$

(iv) $4^{-3} = (4)^{-3} = (2^2)^{-3} = 2^{2 \times -3} = 2^{-6}$

Exercise 1.13

Express questions 1–20 with a single index.

1. $2^3 \times 2^4$
2. $5^2 \times 5^7$
3. $7^3 \times 7$
4. $3 \times 3^4 \times 3^2$
5. $\dfrac{3^7}{3^5}$
6. $\dfrac{2^8}{2^5}$
7. $\dfrac{3^4}{3^6}$
8. $\dfrac{5}{5^4}$
9. $(3^2)^4$
10. $(5^3)^2$
11. $(\sqrt{5})^4$
12. $(\sqrt{3})^3$
13. $\dfrac{2^2}{\sqrt{2}}$
14. $\dfrac{3^4}{\sqrt{3}}$
15. $\dfrac{\sqrt{2}}{2^3}$
16. $\dfrac{\sqrt{5}}{5^2}$
17. $a^{2/3} \times a^{4/3}$
18. $a^{1/2} \times a^{1/2} \times a^2$
19. $\dfrac{a^{7/2}}{a^{3/2}}$
20. $\dfrac{a^2 \times a^{5/2}}{a^{1/2}}$

Express questions 21–32 in the form a^p, where $a \in \mathbb{N}$ and $p \neq 1$.

21. 4
22. 25
23. 36
24. 27
25. 16
26. 49
27. 32
28. 81
29. 125
30. 128
31. 243
32. 216

Write questions 33–62 without indices.

33. 2^3
34. 3^2
35. 4^3
36. 5^2
37. 6^2
38. 5^3
39. 3^4
40. 5^3
41. 7^2
42. 2^4
43. 6^3
44. 8^0
45. 3^{-1}
46. 4^{-2}
47. 5^{-3}
48. 2^{-5}
49. 3^{-2}
50. 10^{-3}
51. $9^{1/2}$
52. $25^{1/2}$
53. $8^{1/3}$
54. $64^{1/3}$
55. $32^{1/5}$
56. $216^{1/3}$
57. $4^{3/2}$
58. $4^{-3/2}$
59. $8^{4/3}$
60. $8^{-4/3}$
61. $27^{2/3}$
62. $27^{-2/3}$

63. Express 64 in the form a^b in three different ways, where $a \in \mathbb{N}$ and $b \neq 1$.
64. Express 81 in the form a^b in two different ways, where $a \in \mathbb{N}$ and $b \neq 1$.

Express questions 65–88 in the form 2^n or 3^n or 5^n or 7^n.

65. 8
66. 9
67. 25
68. 49
69. 16
70. 125
71. $\sqrt{2}$
72. $\sqrt{3}$
73. $\sqrt{5}$
74. $\sqrt{7}$
75. $2\sqrt{2}$
76. $(2\sqrt{2})^2$
77. $\dfrac{4}{\sqrt{2}}$
78. $\dfrac{9}{\sqrt{3}}$
79. $\dfrac{25}{\sqrt{5}}$
80. $\sqrt{125}$
81. $\dfrac{1}{\sqrt{7}}$
82. $\dfrac{125}{\sqrt{5}}$
83. $\left(\dfrac{4}{\sqrt{2}}\right)^2$
84. $\left(\dfrac{1}{\sqrt{3}}\right)^2$
85. $\left(\dfrac{\sqrt{5}}{25}\right)^2$
86. $\left(\dfrac{\sqrt{8}}{2}\right)^2$
87. $\left(\dfrac{25}{\sqrt{5}}\right)^2$
88. $\left(\dfrac{\sqrt{3}}{9}\right)^2$

89. Show that (i) $\left(\dfrac{\sqrt{a}}{a^2}\right)^2 = a^{-3}$ (ii) $\dfrac{(a\sqrt{a})^3}{a^4} = \sqrt{a}$

90. Two of the numbers $9^{1/2}$, 9^{-1}, 27^0, $(-3)^2$ and 3^{-2} are equal. Write down these two values. Justify your answer.

91. The foundation for a building is in the shape of the letter L, as shown. The shape is formed from two squares of dimensions x m and $\sqrt{x}$ m.

    (i) Write down an expression in terms of x of the area of the foundation.

    (ii) If the area of the foundation is 42 m^2, write a quadratic equation in terms of x and calculate the value of x.

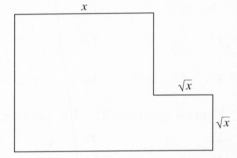

92. (i) Simplify $(x + \sqrt{x})(x - \sqrt{x})$ when $x > 0$.

    (ii) Hence or otherwise, find the value of x for which $(x + \sqrt{x})(x - \sqrt{x}) = 6$.

93. (i) Simplify: (a) $\sqrt{a^2}$ (b) $\sqrt{b^2}$ (c) $\sqrt{x^2}$ (d) $\sqrt{(x+3)^2}$

    (ii) Factorise: (a) $x^2 + 2x + 1$ (b) $x^2 + 4x + 4$

    (iii) Simplify $\sqrt{x^2 + 4x + 4} + \sqrt{x^2 + 2x + 1}$, given that $x \geq 0$.

    (iv) Given that $x \geq 0$, solve for x, $\sqrt{x^2 + 4x + 4} + \sqrt{x^2 + 2x + 1} = x^2$.

94. Express the following in the form $x^{a/b}$.

    (i) $\sqrt{x}$ (ii) $\sqrt[3]{x}$ (iii) $\sqrt[3]{x^2}$ (iv) $\sqrt[4]{x^3}$

ALGEBRA

Exponential equations

Exponent is another name for power or index.
An equation involving the variable in the power is called an **exponential equation**.
For example, $3^{2x+3} = 9$ is an exponential equation.

Exponential equations are solved with the following steps.

1. Write all the numbers as powers of the same number (usually a prime number).
2. Write both sides as one power of the same number using the laws of indices.
3. Equate these powers and solve the equation.

EXAMPLE

Find the value of x if: (i) $4^{x+1} = 128$ (ii) $5^{3x+1} = \dfrac{125}{\sqrt{5}}$

Solution:

(i) 1. $4^{x+1} = 128$ (both 4 and 128 can be written as powers of 2)
$(2^2)^{x+1} = 2^7$ ($4 = 2^2$ and $128 = 2^7$)
2. $2^{2(x+1)} = 2^7$ (multiply the indices on the left-hand side)
$2^{2x+2} = 2^7$ ($2(x+1) = 2x + 2$)
3. $2x + 2 = 7$ (equate the powers)
$2x = 5$
$x = \dfrac{5}{2}$

(ii) 1. $5^{3x+1} = \dfrac{125}{\sqrt{5}}$ (both 125 and $\sqrt{5}$ can be written as powers of 5)
$5^{3x+1} = \dfrac{5^3}{5^{1/2}}$ ($125 = 5^3$ and $\sqrt{5} = 5^{1/2}$)
2. $5^{3x+1} = 5^{3-\frac{1}{2}}$ (subtract index on the bottom from the index on top)
$5^{3x+1} = 5^{2\frac{1}{2}}$ ($3 - \frac{1}{2} = 2\frac{1}{2}$)
3. $3x + 1 = 2\frac{1}{2}$ (equate the powers)
$6x + 2 = 5$ (multiply both sides by 2)
$6x = 3$
$x = \dfrac{1}{2}$

Exercise 1.14
Solve questions 1–24 for x.

1. $5^{2x} = 5^8$
2. $3^{2x+1} = 3^7$
3. $2^{3x-1} = 2^5$
4. $7^{5x-1} = 7^4$
5. $2^x = 4$
6. $3^{x+1} = 9$
7. $2^{x-2} = 16$
8. $3^{2x-1} = 27$
9. $9^{x+1} = 81$
10. $4^{2x-5} = 64$
11. $5^{x+1} = 125$
12. $7^{3x-1} = 49$
13. $4^x = 8$
14. $9^{2x} = 27$
15. $16^{x+1} = 32$
16. $49^x = 7^{2+x}$
17. $3^x = \dfrac{1}{9}$
18. $2^x = \dfrac{1}{8}$
19. $7^x = \dfrac{1}{49}$
20. $5^{2x-1} = \dfrac{1}{125}$
21. $9^{x+1} = \dfrac{1}{27}$
22. $4^{x-3} = \dfrac{1}{32}$
23. $25^{x-2} = \dfrac{1}{125}$
24. $8 \times 2^x = \dfrac{1}{128}$

25. Express $32^{\frac{4}{5}}$ in the form 4^n. Hence or otherwise, solve $4^{2x-1} = 32^{\frac{4}{5}}$.

26. Express $2^5 - 2^4$ in the form 2^n. Hence, solve $2^{3x-5} = 2^5 - 2^4$.

27. Find the two values of x for which $\dfrac{2^{x^2}}{2^x} = 4$.

28. Express $\dfrac{4}{\sqrt{2}}$ in the form $2^{a/b}$. Hence, solve for x, $2^{2x+1} = \dfrac{4}{\sqrt{2}}$.

29. (i) Write each of the following as a power of 2.

   (a) $\sqrt{2}$ (b) 8 (c) $8^{4/3}$ (d) $8\sqrt{2}$

   (ii) Hence, solve for x.

   (a) $2^{5x-1} = 8^{4/3}$ (b) $2^{5x-4} = 8\sqrt{2}$ (c) $2^{3x+1} = (8\sqrt{2})^2$

30. (i) Write each of the following as a power of 3.

   (a) $\sqrt{3}$ (b) $\dfrac{1}{\sqrt{3}}$ (c) 81 (d) $\dfrac{81}{\sqrt{3}}$

   (ii) Hence, solve for x.

   (a) $3^{2x} = \sqrt{3}$ (b) $3^{2x-1} = \dfrac{1}{\sqrt{3}}$ (c) $3^{x-2} = \dfrac{81}{\sqrt{3}}$

31. (i) Write each of the following as a power of 5.

   (a) 125 (b) $\sqrt{5}$ (c) $\dfrac{125}{\sqrt{5}}$ (d) $\left(\dfrac{125}{\sqrt{5}}\right)^2$

   (ii) Hence, solve for x.

   (a) $5^{3x+1} = \sqrt{5}$ (b) $5^{2x+3} = \dfrac{125}{\sqrt{5}}$ (c) $5^{2x-1} = \left(\dfrac{125}{\sqrt{5}}\right)^2$

32. (i) Simplify $\sqrt{x}\left(\sqrt{x} + \dfrac{1}{\sqrt{x}}\right)$, where $x > 0$.

   (ii) Hence or otherwise, solve for x, $\sqrt{x}\left(\sqrt{x} + \dfrac{1}{\sqrt{x}}\right) = 5$ and verify your answer.

33. (i) Write down the formula for the area of a triangle.

(ii) The triangle shown has a base length of $2x\sqrt{x}$ cm and perpendicular height of $\sqrt{x}$ cm.

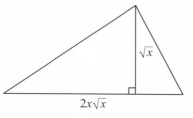

   (a) Express its area, A, in terms of x.

   (b) If $A = 81$ cm^2, calculate the value of x and verify your answer.

34. The rectangle shown has a length of 4 m and a width of 2^k m.

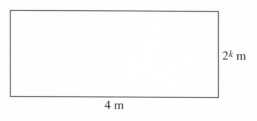

   (i) Express its area, A, in the form 2^{k+p} m^2.

   (ii) Calculate k if:

   (a) $A = 32$ m^2

   (b) $A = 4\sqrt{2}$ m^2

35. In each case, verify your answer.

(i) Express b in terms of a and c where $\dfrac{8a - 5b}{b} = c$.

(ii) Hence or otherwise, evaluate b when $a = 2^{\frac{5}{2}}$ and $c = 3^3$.

CHAPTER 2 ARITHMETIC

Proportion

A proportion of an object is usually given as a fraction comparing a part of the object with the complete object. For example, if a basket of 20 flowers contains 8 roses, then the proportion of the roses in the basket is $\frac{8}{20}$ or $\frac{2}{5}$. Comparing the top with the bottom, we can say that 2 out of every 5 flowers are roses.

A proportion can be given as a percentage. We could say that $\frac{2}{5} = 40\%$ of the flowers are roses.

Note: It is sometimes easier to compare proportions if they are converted to percentages.

EXAMPLE 1

(i) A class of 30 students contains 25 who own a mobile phone. What proportion do not own a mobile phone?

(ii) In an election, the proportion of voters who supported the Happy Party was 55%. There were 12,000 voters in total.
  (a) Express the proportion as a fraction.
  (b) Find the number of Happy Party supporters.

Solution:

(i) There are $30 - 25 = 5$ students who do not own a mobile phone.
The proportion is $\frac{5}{30}$ or $\frac{1}{6}$.

(ii) (a) $55\% = \frac{55}{100} = \frac{11}{20}$

(b) $\frac{11}{20}$ of $12{,}000 = \frac{11}{20} \times 12{,}000$
$= 6{,}600$ supporters.

ARITHMETIC

EXAMPLE 2

A packet of 12 sweets contains three flavours: strawberry, lemon and orange. There are four strawberry and six lemon sweets. The rest are orange flavoured.

(i) How many are orange?

(ii) Find the proportions of each flavour.

(iii) Show that the total of the proportions add up to 1.

(iv) If the flavours are supposed to be equal, what proportion should there be of each flavour?

Solution:

(i) The are 12 sweets in the packet, so:

Orange = 12 − (Strawberry + Lemon) = 12 − (4 + 6) = 2

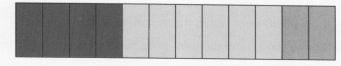

(ii) Strawberry: $\frac{4}{12} = \frac{1}{3}$;   Lemon: $\frac{6}{12} = \frac{1}{2}$;   Orange: $\frac{2}{12} = \frac{1}{6}$.

(iii) Using the unsimplified fractions: $\frac{4}{12} + \frac{6}{12} + \frac{2}{12} = \frac{12}{12} = 1$.

(iv) 12 sweets, three flavours, so ideally 12 ÷ 3 = 4 sweets per flavour.

Proportion: $\frac{4}{12} = \frac{1}{3}$.

OR

Three flavours so $\frac{1}{3}$ per flavour would be expected.

Exercise 2.1

1. There are two rotten apples at the bottom of a barrel containing 12 apples. Express the proportion of rotten apples as a (simplified) fraction and as an exact percentage.

2. A prize fund of €1,200 is distributed as follows: $\frac{1}{2}$ for first prize, $\frac{2}{5}$ for second prize and the remainder for third prize.

   (i) Calculate the proportion of the third prize.   (ii) How much is the third prize worth?

3. In English text, the proportions of the most frequent letters are: **e** 13%, **a** 8% and **t** 9%.
   (i) Rene Descrates once said:

   I think; therefore I am.

   (a) Ignoring punctuation and spaces, write down the proportions of the letters **e**, **a** and **t** in the above quote.

   (b) For each letter, indicate whether it is more or less frequent than usual.

   (ii) Isaac Newton once said:

   If I have seen further than others, it is by standing upon the shoulders of giants.

   (a) Ignoring punctuation and spaces, write down the proportions of the letters **e**, **a** and **t** in the above quote.

   (b) For each letter, indicate whether it is more or less frequent than usual.

   (iii) Why would the proportions be so different?

4. In a survey, the proportion of people who watched a football match on television was $\frac{3}{8}$.
   (i) What proportion did not watch the match?
   (ii) If 240,000 people did not watch the match, how many did?

5. €400 is $\frac{8}{11}$ of a prize fund. Find the total prize fund.

6. When a cyclist had travelled a distance of 17 km, she had completed $\frac{5}{8}$ of her journey. What was the length of the journey?

Dividing quantities in a given ratio

Ratios can be used to divide, or share, quantities.

To divide, or share, a quantity in a given ratio, do the following:

> 1. Add the ratios to get the total number of parts.
> 2. Divide the quantity by the total of the parts (this gives one part).
> 3. Multiply this separately by each ratio to find the shares.

Check your answers by confirming that the shares add up to the original total.

ARITHMETIC

EXAMPLE 1

Divide: (i) €450 in the ratio 5 : 3 : 7 (ii) 64 kg in the ratio $1 : \frac{1}{3} : 4$

Solution:

(i) Number of parts = 5 + 3 + 7 = 15

$$1 \text{ part} = \frac{€450}{15} = €30$$

5 parts = €30 × 5 = €150
3 parts = €30 × 3 = €90
7 parts = €30 × 7 = €210

∴ €450 in the ratio 5 : 3 : 7
= €150, €90, €210

Check:
€150 + €90 + €210 = €450 ✓

(ii) $1 : \frac{1}{3} : 4 = 3 : 1 : 12$

(multiply each part by 3)
Number of parts = 3 + 1 + 12 = 16

$$1 \text{ part} = \frac{64 \text{ kg}}{16} = 4 \text{ kg}$$

3 parts = 4 kg × 3 = 12 kg
12 parts = 4 kg × 12 = 48 kg

∴ 64 kg in the ratio $1 : \frac{1}{3} : 4$

= 12 kg, 4 kg, 48 kg

Check:
12 kg + 4 kg + 48 kg = 64 kg ✓

Sometimes we are given an equation in disguise.

EXAMPLE 2

Amy and Beatrice share a prize in the ratio 7 : 5.
Beatrice gets €45. How much should Amy get?

Solution:
Equation given in disguise:

5 parts = €45   (Beatrice's share = 5 parts)

$$1 \text{ part} = \frac{€45}{5} = €9$$ (divide both sides by 5)

Amy's share = 7 parts = 7 × €9 = €63.

Exercise 2.2

1. Divide: (i) €80 in the ratio 7 : 3 (ii) 450 g in the ratio 5 : 4
2. Divide: (i) €480 in the ratio 3 : 4 : 5 (ii) €4,000 in the ratio 5 : 8 : 7
3. Divide: (i) 238 g in the ratio 7 : 2 : 5 (ii) 162 cm in the ratio 4 : 3 : 2
4. Divide: (i) €504 in the ratio 3 : 4 : 5 (ii) 336 cm in the ratio 2 : 5 : 7
5. Divide: (i) €374 in the ratio 6 : 7 : 9 (ii) 1,560 g in the ratio 8 : 1 : 4

In questions 6–9, write the given ratio as a ratio of whole numbers, in its simplest form.

6. $1\frac{1}{2} : 2$
7. $\frac{1}{3} : 1$
8. $\frac{1}{2} : \frac{1}{4} : 1$
9. $\frac{3}{4} : \frac{1}{2} : 1$

10. Divide: (i) €42 in the ratio $1 : \frac{1}{2}$ (ii) 280 g in the ratio $\frac{1}{2} : 2$
11. Divide: (i) €210 in the ratio $1 : 2 : \frac{1}{2}$ (ii) 585 cm in the ratio $\frac{1}{2} : 2 : 5$
12. Divide: (i) 546 g in the ratio $1 : \frac{2}{3} : \frac{1}{2}$ (ii) €920 in the ratio $\frac{1}{2} : \frac{2}{3} : \frac{3}{4}$

13. In a competition, team A scored $22\frac{1}{2}$ points and team B scored $17\frac{1}{2}$ points. The two teams share a prize of €28,000 in proportion to the number of points they scored. How much money does each team receive?

14. One town, A, has a population of 4,800 and a second town, B, has a population of 6,720. The two towns share a grant of €429,120 in proportion to their populations. How much does town A receive?

15. (i) David, Eric and Fred decide to buy a lottery ticket for €2. David pays 50c, Eric pays 90c and Fred pays the remainder. In what ratio should a prize be shared?
    (ii) If their ticket wins a prize of €500, how much should each get?

16. A and B share a sum of money in the ratio 2 : 3. If A's share is €80, calculate B's share.

17. Two lengths are in the ratio 7 : 5. If the larger length is 140 cm, calculate the other length.

18. The lengths of the sides of a triangle are in the ratio 4 : 3 : 2. If the shortest side is of length 36 cm, calculate the perimeter of the triangle.

19. P, Q and R share a sum of money in the ratio 2 : 4 : 5, respectively. If Q's share is €60, find: (i) P's share (ii) the total sum of money shared.

20. The profits of a business owned by A, B and C are shared in the ratio of their investments, €32,000, €16,000 and €20,000, respectively. If C received €6,350, how much did A receive?

21. A glass rod falls and breaks into three pieces whose lengths are in the ratio 8 : 9 : 5. If the sum of the lengths of the two larger pieces is 119 cm, find the length of the third piece.

22. A woman gave some money to her four children in the ratio 2 : 3 : 5 : 9. If the difference between the largest and the smallest share is €3,500, how much money did she give altogether?

23. €360 is divided between A and B in the ratio 3 : k. If A received €135, find the value of k.

24. Roy and Sam share €440 in the ratio 8 : 3.

  (i) How much does each get?

  (ii) If Roy gives €45 of his share to Sam, in what ratio is the money now?

Percentages

In many questions dealing with percentages, we will not be given the original amount. The best way to tackle this type of problem is to treat it as an equation given in disguise. From this we can find 1% and, hence, any percentage we like.

EXAMPLE 1

An auctioneer's fee for the sale of a house is $1\frac{1}{2}\%$ of the selling price.

If the fee is €4,200, calculate the selling price.

Solution:
Given: auctioneer's fee is €4,200.

$1\frac{1}{2}\% = €4,200$ (equation given in disguise)

$3\% = €8,400$ (multiply both sides by 2)

$1\% = €2,800$ (divide both sides by 3)

$100\% = €280,000$ (multiply both sides by 100)

∴ The selling price of the house was €280,000.

EXAMPLE 2

A bill for €102·85 includes VAT at 21%. Calculate the amount of the bill before VAT is added.

Solution:
Think of the bill before VAT is added on as 100%. After 21% VAT, this becomes 121%.

Given: 121% = €102·85   (equation given in disguise)
         1% = €0·85   (divide both sides by 121)
         100% = €85   (multiply both sides by 100)

∴ The bill before VAT is added is €85.

EXAMPLE 3

A shopkeeper buys an armchair for €410. He adds a mark-up of 40% to find his basic sale price. VAT must be added at a rate of 21%. Suggest a suitable retail price point.

Solution:
To add 40%, we find 140% of the amount.

As $140\% = \dfrac{140}{100} = 1\cdot 4$, a simple way to do this is to multiply by 1·4.

€410   (base price)
€410 × 1·4 = €574   (add 40% mark-up)
€574 × 1·21 = €694.54   (add 21% VAT)

A suitable price point might be €695, €699 or €699·99.

Exercise 2.3

1. Calculate: (i) 8% of €120   (ii) 12% of €216   (iii) 21% of €124

2. A musical store owner pays €2,800 for a set of drums and marks it up so that he makes a profit of 35%. Then 21% VAT is added on. Calculate the selling price.

3. (i) A shopkeeper pays €240 for a bicycle and marks it up so that she makes a profit of 20%. Find the selling price.

   (ii) During a sale, the price of the bicycle is reduced by 15%.
   Calculate: (a) the sale price (b) the percentage profit on the bicycle during the sale

4. One litre of water is added to four litres of milk in a container. Calculate the percentage of water in the container.

5. A tank contains 320 litres of petrol. 128 litres are removed. What percentage of the petrol remains in the tank?

6. A bill for €96·76 includes VAT at 18%. Calculate the amount of the bill before VAT is added.

7. A bill for €58·08 includes VAT at 21%. Calculate the amount of VAT in the bill.

8. When a woman bought a television set in a shop, VAT at 21% was added on.
   (i) If the VAT on the cost of the set was €252, what was the price of the television set before VAT was added?
   (ii) What was the price including VAT?

9. A boy bought a calculator for €74·75, which included VAT at 15%. Find the price of the calculator if VAT was reduced to 12%.

10. When the rate of VAT was increased from 18% to 21%, the price of a guitar increased by €81. Calculate the price of the guitar, inclusive of the VAT at 21%.

11. When 9% of the pupils in a school are absent, 637 are present. How many pupils are on the school roll?

12. 15% of a number is 96. Calculate 25% of the number.

13. A salesperson's commission for selling a car is $2\frac{1}{2}$% of the selling price. If the commission for selling a car was €350, calculate the selling price.

14. A solicitor's fee for the sale of a house is $1\frac{1}{2}$% of the selling price. If the fee is €3,480, calculate the selling price.

15. In a sale, the price of a piece of furniture was reduced by 20%. The sale price was €1,248. What was the price before the sale?

16. A salesperson's income for a year was €59,000. This was made up of basic pay of €45,000 plus a commission of 4% of sales. Calculate the amount of the sales for the year.

17. A fuel mixture consists of 93% petrol and 7% oil. If the mixture contains 37·2 litres of petrol, calculate the volume of oil.

18. A book of raffle tickets sells for €10. The prizes in euro are 2,000, 1,000, 500 and 250. If printing costs amount to €250, calculate the smallest number of books which must be sold to: **(i)** Cover costs   **(ii)** Make a profit of €3,000

19. A lottery had a first prize of 70% of the prize fund and a consolation prize of 30%. Six people shared the first prize and each received €3,500. If the consolation prize was divided between 450 people, how much did each receive?

20. **(i)** A tanker delivered heating oil to a school. Before the delivery, the meter reading showed 11,360 litres of oil in the tanker. After the delivery, the meter reading was 7,160 litres. Calculate the cost of the oil delivered if 1 litre of oil cost 36·5c.

    (ii) When VAT was added to the cost of the oil delivered, the bill to the school amounted to €1,808·94. Calculate the rate of VAT added.

21. **(i)** An antiques dealer bought three chairs at an auction. He sold them later for €301·60, making a profit of 16% on their total cost. Calculate the total cost of the chairs.

    (ii) The first chair cost €72 and it was sold at a profit of 15%. Calculate its selling price.

    (iii) The second chair cost €98 and it was sold for €91. Find the percentage profit made on the sale of the third chair.

22. A retailer buys an item for €299. Can she add on a 40% mark-up and still sell it for under €500 including 21% VAT?

23. To match a competitor's price of €14,950 (including 21% VAT), a car dealer has to recalculate his potential profit. He purchased the car for €9,200.

    The dealer usually applies a mark-up of between 30% and 35% – can he match the price or make an even better offer to the customer?

24. A shopkeeper want to sell toys at three prices: €9·99, €12·99 and €24·99, including 21% VAT. What is the maximum price that the shopkeeper must pay for each item and still include a minimum of a 30% mark-up? Give your answers correct to the nearest cent.

25. During 1874, the population of a small town rose by 4%. The following year, the population declined by 4%. Would the population at the start of 1,876 be less than, equal to or greater than the population two years earlier? Explain your answer.

Relative error and percentage error

When calculations are being made, errors can occur, especially calculations which involve rounding. It is important to have a measure of the error.

Definitions

> Error = | true value − estimate value | and is always considered positive.

> Relative error = $\dfrac{\text{Error}}{\text{True value}}$

> Percentage error = $\dfrac{\text{Error}}{\text{True value}} \times 100\%$

EXAMPLE 1

A distance of 190 km was estimated to be 200 km. Calculate:
(i) The error **(ii)** The relative error **(iii)** The percentage error, correct to one decimal place

Solution:
True value = 190 km.   Estimated value = 200 km.

(i) Error = |true value − estimated value| = |190 − 200| = |−10| = 10 km (positive value)

(ii) Relative error = $\dfrac{\text{Error}}{\text{True value}} = \dfrac{10}{190} = \dfrac{1}{19}$

(iii) Percentage error = $\dfrac{\text{Error}}{\text{True value}} \times 100\% = \dfrac{10}{190} \times 100\% = 5 \cdot 3\%$ (correct to one decimal place)

EXAMPLE 2

The answer to 5·6 + 7·1 was given as 12·5.

What was the percentage error, correct to two decimal places?

Solution:
True value = 5·6 + 7·1 = 12·7.   Estimated value = 12·5.
Error = |true value − estimated value| = |12·7 − 12·5| = |0·2| = 0·2

Percentage error = $\dfrac{\text{Error}}{\text{True value}} \times 100\% = \dfrac{0 \cdot 2}{12 \cdot 7} \times 100\% = 1 \cdot 57\%$ (correct to two decimal places)

Exercise 2.4
Complete the following table.

	True value	Estimated value	Error	Relative error (as a fraction)	Percentage error (correct to one decimal place)
1.	12	10			
2.	43	40			
3.	136	140			
4.	4·8	5			
5.	5·7	6			
6.	390	400			

7. The depth of a swimming pool was estimated to be 1·5 m. The true depth was 1·65 m. Find: **(i)** The error **(ii)** The percentage error, correct to one decimal place

8. The mass of a rock is estimated to be 65 kg. Its true mass is 67·5 kg. Find: **(i)** The error **(ii)** The percentage error, correct to one decimal place

9. A distance of 105 km was estimated to be 100 km. Calculate:
   (i) The error **(ii)** The relative error **(iii)** The percentage error, correct to one decimal place

10. The estimate for building a wall was €3,325. The actual cost was €3,500. Calculate the percentage error.

11. The number of people estimated to be at a meeting was 400. The actual number who attended the meeting was 423. Calculate the percentage error, correct to two decimal places.

12. The value of $\dfrac{48\cdot27 + 12\cdot146}{14\cdot82 - 3\cdot02}$ was estimated to be 5.

    Calculate the percentage error, correct to one decimal place.

13. The value of $\dfrac{30\cdot317}{\sqrt{24\cdot7009}}$ was estimated to be 6.

    Calculate: **(i)** The error **(ii)** The percentage error, correct to two decimal places

14. **(i)** Calculate the volume of a solid cylinder of radius 6 cm and height 14 cm $\left(\text{assume } \pi = \tfrac{22}{7}\right)$.
    (ii) When doing this calculation, a student used $\pi = 3$. Calculate the student's percentage error in the calculation, assuming that $\pi = \tfrac{22}{7}$ is the exact value, correct to one decimal place.

15. Calculate the percentage error in calculating the total of 324 + 432 + 234 if the digit 4 is replaced by a 5 each time. Give your answer correct to one decimal place.

16. Four items in a shop cost €7·70, €14·90, €16·80 and €23·10.
    (i) Frank estimates the total cost of the four items by ignoring the cent part in the cost of each item. Calculate the percentage error in his estimate.
    (ii) Fiona estimates the total cost of the four items by rounding the cost of each item to the nearest euro. Calculate the percentage error in her estimate.

17. A food manufacturer sells carrots in 500 g tins. For two hours, a faulty machine filled 12,500 tins with only 495 g of carrots.
    (i) Calculate the percentage error.
    (ii) How many kg of carrots were used during the two hours?
    (iii) How many tins would usually be filled with the quantity used?
    (iv) If the tins had been sold at the usual price of 50c, how much profit would the manufacturer have made because of the fault?

Tolerance

Modern machines are manufactured from hundreds of components which need to fit together accurately. For that reason, the various parts must be made to exact measurements, although the level of exactness can vary. For example, the parts which make up a cruise ship might not need to be manufactured as accurately as parts of a watch.

Objects which must be made very accurately (such as a watch) would have a very low tolerance value. For example, a small circular cog for a watch might need to be 8 mm in diameter. If it was 10 mm in diameter, it would be unlikely to fit. The watchmaker could describe the part as needing a diameter of 8 mm ± 0·5 mm.

This would mean that the smallest acceptable diameter is 8 − 0·5 mm = 7·5 mm and the largest acceptable diameter is 8 + 0·5 mm = 8·5 mm.

EXAMPLE

A shipbuilder needs sheets of metal to be 2 m long by 1 m wide but will accept sheets which are 2 m ± 3 mm by 1 m ± 2 mm.

(i) What are the dimensions (in mm) of the largest sheets and the smallest within the given tolerances?

(ii) Calculate, in m^2, the difference between the area of the largest and the smallest sheets.

Solution:
Draw a sketch:

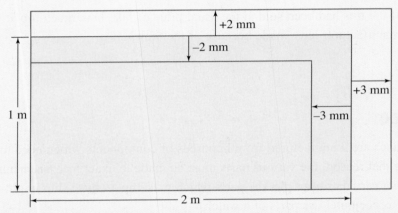

The shaded part shows the desired sheet. The larger outline shows the maximum sheet, while the smaller outline shows the minimum.

(i) Maximum length: 2 m + 3 mm = 2,000 mm + 3 mm = 2,003 mm
Maximum width: 1 m + 2 mm = 1,000 mm + 4 mm = 1,002 mm
Maximum dimensions: 2,003 mm × 1,002 mm

Minimum length: 2 m − 3 mm = 2,000 mm − 3 mm = 1,997 mm
Minimum width: 1 m − 2 mm = 1,000 mm − 4 mm = 998 mm
Minimum dimensions: 1,997 mm × 998 mm

(ii) Convert: 2,003 mm = 2·003 m;   1,002 mm = 1·002 m
Maximum area: 2·003 m × 1·002 m = 2·007 006 m^2
Convert: 1,997 mm = 1·997 m;   998 mm = 0·998 m
Maximum area: 1·997 m × 0·998 m = 1·993 006 m^2

∴ The difference in the areas = (2·007 006 − 1·993 006) m^2 = 0·014 m^2

ARITHMETIC

Exercise 2.5

In questions 1–6, calculate the minimum and maximum values, using appropriate units.

	Measurement	Tolerance	Minimum value	Maximum value
1.	12	±2		
2.	120	±5		
3.	80%	±5%		
4.	4 m	±5 mm		
5.	€3,200	±€450		
6.	3	±5		

7. A packet of sweets describes the contents as containing 150 ± 2% sweets. What is the tolerance in terms of sweets?

8. A bridge over a motorway needs to span a distance of 40 m. After some consultation, the specification for the bridge requires a width of 39·9 ± 0·1 m.
   - (i) By how much can the bridge vary?
   - (ii) Why would the bridge need to be shorter than the distance it needs to span?

9. A sheet of metal is manufactured to be 3 m by 1 m and 1 cm thick.
   - (i) What is its volume in cm^3?
   - (ii) The manufacturing process means that the thickness is highly accurate but the length and width are only accurate to ± 1 cm. What are the minimum and maximum volumes?
   - (iii) The sheet is made of steel, which weighs 8 g per cm^3. What is the difference in weight between the smallest and the largest sheet?
   - (iv) Given the difference in volume and also in weight, discuss the consequences of this tolerance.

Foreign exchange

Currency is another name for money. In the European Union the unit of currency is called the euro (€). The method of direct proportion is used to convert one currency into another currency.

Note: Write down the equation given in disguise, putting the currency we want to find on the right-hand side.

EXAMPLE 1

A book cost €8·60 in Dublin and US$10·08 in New York. If €1 = US$1·20, in which city is the book cheaper and by how many European cent?

Solution:
Express the price of the book in New York in euro and compare to the price in Dublin.

$$US\$1·20 = €1 \quad \text{(€ on the right because we want our answer in €)}$$

$$US\$1 = €\frac{1}{1·20} \quad \text{(divide both sides by 1·20)}$$

$$\$10·08 = €\frac{1}{1·20} \times 10·08 \quad \text{(multiply both sides by 10·08)}$$

$$\$10.08 = €8·40 \quad \text{(simplify the right-hand side)}$$

Difference = €8·60 − €8·40 = 20c

∴ The book is cheaper in New York by 20c.

EXAMPLE 2

A person changes €500 into Japanese yen (¥). A charge is made for this transaction. The exchange rate is €1 = ¥320. If the person receives ¥156,000, calculate the percentage charge.

Solution:
Express €500 in ¥.

$$€1 = ¥320 \quad \text{(¥ on the right because we want our answer in ¥)}$$

$$€500 = ¥500 \times 320 \quad \text{(multiply both sides by 500)}$$

$$€500 = ¥160,000 \quad \text{(full amount due)}$$

Amount received = ¥156,000

∴ Charge = ¥160,000 − ¥156,000 = ¥4,000

$$\text{Percentage charge} = \frac{\text{Charge}}{\text{Full amount}} \times 100\% = \frac{4,000}{160,000} \times 100\% = 2·5\%.$$

Exercise 2.6

1. If €1 = $1·03, find the value of: (i) €250 in dollars (ii) $618 in euro
2. A train ticket costs $54. If €1 = $1·08, calculate the cost of the ticket in euro
3. €1 = ¥304 (Japanese yen).
    (i) How many yen would you receive for €240?
    (ii) How many euro would you receive for ¥115,520?
4. A part for a tractor costs €600 in France and the same part costs R1,368 in South Africa. If €1 = R2·4, in which country is it cheaper, and by how much (in euro)?
5. (i) A tourist changed €5,000 on board a ship into South African rand at a rate of €1 = R2·2. How many rand did she receive?
    (ii) When she came ashore she found that the rate was €1 = R2·35. How much did she lose, in rand, by not changing her money ashore?
6. (i) A person buys 167,400 Japanese yen when the exchange rate is €1 = ¥310. A charge is made for this. How much, in euro, is this charge if the person pays €548·10?
    (ii) Calculate the percentage commission on the transaction.
7. When the exchange rate is €1 = $0·98, a person buys 3,430 dollars from a bank. If the bank charges a commission of $2\frac{1}{2}\%$, calculate the total cost in euro.
8. (i) A person buys 5,160 Canadian dollars when the exchange rate is €1 = $2·15. A charge (commission) is made for this service. How much, in euro, is this charge if the person pays €2,448?
    (ii) Calculate the percentage commission on the transaction.
9. Dollars were bought for €8,000 when the exchange rate was €1 = $1·02. A commission was charged for this service. If the person received $7,956, calculate the percentage commission charged.
10. If €1 = $1·10 and €1 = R2·64, how many dollars can be exchanged for 2,112 rand?
11. An importer buys goods for $1,286·40 when the exchange rate is €1 = $1·34. If he sells them for €1,100, find his profit in euro.
12. A supplier agrees to buy 100 computers for $600 each. She plans to sell them for a total of €62,400.
    (i) Calculate the percentage profit, on the cost price, she will make if the exchange rate is €1 = $1·25.
    (ii) Calculate the percentage profit, on the cost price, if the exchange rate changes to €1 = $1·20.

Interest

Interest is the sum of money that you pay for borrowing money or that is paid to you for lending money. When dealing with interest, we use the following symbols:

> P = the **principal**, the sum of money borrowed or invested at the beginning of the period.
> t = the **time**, the number of weeks/months/years for which the sum of money is borrowed or invested.
> i = the **interest rate**, the percentage rate per week/month/year expressed as a fraction or a decimal at which interest is charged.
> A = the **amount**, the amount of money, including interest, at the end of a week/month/year.
> F = the **final amount**, the final sum of money, including interest, at the end of the period.

Note: per annum = per year.

Compound interest

When a sum of money earns interest, this interest is often added to the principal to form a new principal. This new principal earns interest in the next year and so on. This is called **compound interest**.

When calculating compound interest, do the following.

Method 1:

> Calculate the interest for the **first** year and add this to the principal to form the new principal for the next year. Calculate the interest for **one** year on this new principal and add it on to form the principal for the next year, and so on. The easiest way to calculate each stage is to multiply the principal at the beginning of each year by the factor:
> $$(1 + i)$$
> This will give the principal for the next year, and so on.

Method 2:
If the number of years is greater than three, then using a formula and a calculator will be much quicker.

> Use the formula: $F = P(1 + i)^t$

Note: The formula does not work if:

> the interest rate, i, is changed during the period
> **or**
> money is added or subtracted during the period.

EXAMPLE 1

Calculate the compound interest on €10,000 for three years at 4% per annum.

Solution:

$$1 + i = 1 + \frac{4}{100} = 1 + 0.04 = 1.04$$

Method 1:

$P_1 = 10,000$ (principal for the first year)
$A_1 = 10,000 \times 1.04 = 10,400$ (amount at the end of the first year)
$P_2 = 10,400$ (principal for the second year)
$A_2 = 10,400 \times 1.04 = 10,816$ (amount at the end of the second year)
$P_3 = 10,816$ (principal for the third year)
$A_3 = 10,816 \times 1.04 = 11,248.64$ (amount at the end of the third year)

Compound interest $= A_3 - P_1 =$ €11,248·64 − €10,000 = €1,248·64.

The working can also be shown using a table:

Year	Principal	Amount
1	10,000	$10,000 \times 1.04 = 10,400$
2	10,400	$10,400 \times 1.04 = 10,816$
3	10,816	$10,816 \times 1.04 = 11,248.64$

Compound interest $= A_3 - P_1 =$ €11,248·64 − €10,000 = €1,248·64.

Method 2:

Given: $P = 10,000$, $i = \frac{4}{100} = 0.04$, $t = 3$. Find F.

$$F = P(1 + i)^t$$
$$= 10,000(1.04)^3$$
$$F = 11,248.64$$

Compound interest $= F - P =$ €11,248·64 − €10,000 = €1,248·64.

EXAMPLE 2

€8,500 was invested for three years at compound interest. The rate for the first year was 6%, the rate for the second year was 8% and the rate for the third year was 5%.

Calculate: **(i)** the amount **(ii)** the compound interest at the end of the third year.

Solution:
As the rate changes each year, we cannot use the formula.

Year	Principal	Amount	$(1 + i)$
1	8,500	$8{,}500 \times 1{\cdot}06 = 9{,}010$	$\left(1 + \frac{6}{100} = 1{\cdot}06\right)$
2	9,010	$9{,}010 \times 1{\cdot}08 = 9{,}730{\cdot}80$	$\left(1 + \frac{8}{100} = 1{\cdot}08\right)$
3	9,730·80	$9{,}730{\cdot}80 \times 1{\cdot}05 = 10{,}217{\cdot}34$	$\left(1 + \frac{5}{100} = 1{\cdot}05\right)$

(i) The amount at the end of the third year is €10,217·34.
   Alternatively, in one calculation: $A_3 = €8{,}500 \times 1{\cdot}06 \times 1{\cdot}08 \times 1{\cdot}05 = €10{,}217{\cdot}34$.
(ii) Compound interest $= A_3 - P_1 = €10{,}217{\cdot}34 - €8{,}500 = €1{,}717{\cdot}34$.

Exercise 2.7

In questions 1–8, calculate the compound interest.

1. €12,000 for 2 years at 8% per annum
2. €15,000 for 2 years at 7% per annum
3. €18,000 for 3 years at 5% per annum
4. €25,000 for 3 years at 8% per annum
5. €750 for 3 years at 10% per annum
6. €5,000 for 3 years at 2% per annum
7. €12,400 for 2 years at 6·5% per annum
8. €80,000 for 3 years at 2·5% per annum
9. €4,000 was invested for two years at compound interest. The interest rate for the first year was 4% and for the second was 5%. Calculate the total interest earned.
10. €6,500 was invested for three years at compound interest. The interest rate for the first year was 5%, for the second year 8% and for the third year 12%. Calculate the total interest earned.
11. €7,500 was invested for three years at compound interest. The rate for the first year was 4%, the rate for the second year was 3% and the rate for the third year was $2\frac{1}{2}$%. Calculate the amount after three years.

Repayments/further investments

In some questions, money is repaid at the end of a year or a further investment is made at the beginning of the next year. It is important to remember that in these cases, the **formula does not work**. In the next example, F_1 = further investment at the beginning of the second year, F_2 = a further investment at the beginning of the third year.

EXAMPLE 1

(i) A person invested €40,000 in a building society. The rate of interest for the first year was $3\frac{1}{2}\%$. At the end of the first year the person invested a further €6,000. The rate of interest for the second year was 4%. Calculate the value of the investment at the end of the second year.

(ii) At the end of the second year a further sum of €4,704 was invested. At the end of the third year the total value of the investment was €55,350. Calculate the rate of interest for the third year.

Solution:

(i) $P_1 = 40{,}000$
$A_1 = 40{,}000 \times 1{\cdot}035$ (amount at the end of the first year)
$ = 41{,}400$
$F_1 = 6{,}000$ (further investment of €6,000)
─────────
$P_2 = 47{,}400$ ($A_1 + F_1 = P_2$ = principal for the second year)
$A_2 = 47{,}400 \times 1{\cdot}04$
$A_2 = 49{,}296$ (amount at the end of the second year)

Therefore, the value of the investment at the end of the second year = €49,296.

(ii) $F_2 = 4{,}704$ (further investment of €4,704)
$P_3 = A_2 + F_2$ ($A_2 + F_2 = P_3$ = principal for the third year)
$P_3 = 49{,}296 + 4{,}704 = 54{,}000$
Given $A_3 = 55{,}350$ (amount at the end of the third year)
Interest for the third year = $A_3 - P_3 = 55{,}350 - 54{,}000 = 1{,}350$

$$\text{Interest rate for the third year} = \frac{\text{Interest for the third year}}{\text{Principal for the third year}} \times 100\%$$

$$= \frac{1{,}350}{54{,}000} \times 100\%$$

$$= 2\tfrac{1}{2}\%$$

EXAMPLE 2

(i) A person invested €20,000 for three years at 6% per annum compound interest. Calculate the amount after two years.

(ii) After two years, a sum of money was withdrawn. The money which remained amounted to €22,260 at the end of the third year. Calculate the amount of money withdrawn after two years.

Solution:

(i)
$P_1 = 20,000$ (principal for the first year)
$A_1 = 20,000 \times 1 \cdot 06 = 21,200$ (amount at the end of the first year)
$P_2 = 21,200$ (principal for the second year)
$A_2 = 21,200 \times 1 \cdot 06 = 22,472$ (amount at the end of the second year)

∴ The amount after two years = €22,472.

(ii) At this point, a sum of money was withdrawn.

What we do is **work backwards** from the end of the third year.

$A_3 = 22,260$ (amount at the end of the third year)
∴ 106% = 22,260 (increased by 6% during the year)
1% = 210 (divide both sides by 106)
100% = 21,000 (multiply both sides by 100)

∴ The principal for the third year, P_3, was €21,000.
But the amount at the end of the second year, A_3, was €22,472.
∴ The sum of money withdrawn at the end of the second year
$= A_2 - P_3 = €22,472 - €21,000 = €1,472$.

Exercise 2.8

1. A woman borrowed €30,000 at 6% per annum compound interest. She agreed to repay €5,000 at the end of the first year, €5,000 at the end of the second year and to clear the debt at the end of the third year. How much was paid to clear the debt?

2. A man borrowed €40,000 at 4% per annum compound interest. He agreed to repay €8,000 at the end of the first year, €9,000 at the end of the second year and to clear the debt at the end of the third year. How much was paid to clear the debt?

3. A man borrowed €15,000. He agreed to repay €2,000 after one year, €3,000 after two years and the balance at the end of the third year. If interest was charged at 8% in the first year,

5% in the second year and 6% in the third year, how much was paid at the end of the third year to clear the debt?

4. €4,000 was invested for one year. The interest earned was €120. Calculate the rate of interest.

5. €2,500 amounts to €2,600 after one year. Calculate the rate of interest.

6. €8,400 amounts to €8,694 after one year. Calculate the rate of interest.

7. €6,500 amounts to €6,662·50 after one year. Calculate the rate of interest.

8. €12,000 was invested for two years at compound interest.
   (i) The interest at the end of the first year was €600. Calculate the rate of interest for the first year.
   (ii) At the end of the second year the investment was worth €13,167. Calculate the rate of interest for the second year.

9. €60,000 is borrowed for two years. Interest for the first year is charged at 4% per annum.
   (i) Calculate the amount owed at the end of the first year.
   (ii) €7,400 is then repaid. Interest is charged at r% per annum for the second year. The amount owed at the end of the second year is €56,650. Calculate the value of r.

10. €40,000 was invested for three years at compound interest. The rate of interest was 4% per annum for the first year and $3\frac{1}{2}$% per annum for the second year.
    (i) Calculate the amount of the investment after two years.
    (ii) A further €2,444 was invested. If the investment amounted to €46,637·50 at the end of the third year, calculate the rate of interest for the third year.

11. (i) A person invested €20,000 in a building society. The rate of interest for the first year was $2\frac{1}{2}$%. At the end of the first year the person invested a further €2,000. The rate of interest for the second year was 2%. Calculate the value of the investment at the end of the second year.
    (ii) At the end of the second year a further sum of €1,050 was invested. At the end of the third year the total value of the investment was €24,720. Calculate the rate of interest for the third year.

12. €10,000 was invested for three years at compound interest. The rate for the first year was 4%. The rate for the second year was $4\frac{1}{2}$%.
    (i) Find the amount of the investment at the end of the second year.
    (ii) At the beginning of the third year a further €8,000 was invested. The rate for the third year was r%. The total investment at the end of the third year was €19,434·04. Calculate the value of r.

13. €75,000 was invested for three years at compound interest. The rate for the first year was 3%. The rate for the second year was $2\frac{1}{2}$%. At the end of the second year €10,681·25 was withdrawn.

   (i) Find the principal for the third year.
   (ii) The rate for the third year was r%. The total investment at the end of the third year was €70,897·50. Calculate the value of r.

In questions 14–17, it may make the working easier to calculate the amount after two years, then work backwards from the end of the third year to find the sum of money withdrawn.

14. A person invested €30,000 for three years at 5% per annum compound interest.

   (i) Calculate the amount after two years.
   (ii) After two years a sum of money was withdrawn. The money which remained amounted to €26,250 at the end of the third year. Calculate the amount of money withdrawn after two years.

15. A person invested €50,000 for three years at 4% per annum compound interest. At the end of the first year, €11,500 was withdrawn. At the end of the second year, another sum of money was withdrawn. At the end of the third year, the person's investment was worth €36,400. Calculate the amount of money withdrawn after two years.

16. €45,000 was invested for three years at compound interest. The interest rate for the first year was 6% per annum, the interest rate for the second year was 4% per annum and the interest rate for the third year was 3% per annum. At the end of the first year €7,700 was withdrawn. At the end of the second year €w was withdrawn. At the end of the third year the investment was worth €37,080. Find the value of w.

17. €60,000 was invested for three years at compound interest. The interest rate for the first year was $3\frac{1}{2}$% per annum, the interest rate for the second year was $2\frac{1}{2}$% per annum and the interest rate for the third year was 2% per annum. At the end of the first year €4,100 was withdrawn. At the end of the second year €w was withdrawn. At the end of the third year the investment was worth €56,100. Find the value of w.

Depreciation

For depreciation we multiply by the factor $(1 - i)$ for each year.
The formula for depreciation is:

$$F = P(1 - i)^t$$

P is the original value at the beginning of the period and F is the final amount at the end of the period.

EXAMPLE

A machine depreciates at 15% per annum. It was bought for €40,000.
How much is it worth after four years?

Solution:

Given: $P = 40{,}000$, $i = \dfrac{15}{100} = 0{\cdot}15$, $t = 4$. Find F.

$$F = P(1 - i)^t$$
$$= 40{,}000(1 - 0{\cdot}15)^4$$
$$F = 20{,}880{\cdot}25$$

Thus, after four years, the machine is worth €20,880·25.

Exercise 2.9

1. A car depreciates at 15% per annum. It was bought for €30,000. How much is it worth after three years?
2. A machine depreciates at 10% per annum. It was bought for €55,000. How much is it worth after four years?

3. A machine cost €100,000 when new. In the first year it depreciates by 15%. In the second year it depreciates by 8% of its value at the end of the first year. In the third year it depreciates by 5% of its value at the end of the second year.

New price	€100,000
Value after 1 year	
Value after 2 years	
Value after 3 years	

   (i) By completing the table, calculate its value after three years.

   (ii) Calculate its total depreciation after three years.

4. A luxury car costs €50,000 when bought new. Each year it depreciates by 20%.

   (i) Copy and complete the following table.

Years after purchase	0	1	2	3	4
Value of car (€)	50,000	40,000			

   (ii) Plot the results on graph paper, putting the value of the car on the vertical axis.

   (iii) The graph is not linear. Why is this a good representation of the way cars depreciate?

5. A car was bought for €25,000. After one year, it had depreciated in value to €23,000.

   (i) What was the annual percentage rate of depreciation?

   (ii) If the car depreciated at this rate for a further three years, what was it worth at the end of the fourth year? Give your answer correct to the nearest euro.

Annual equivalent rate (AER) and annual percentage rate (APR)

Nowadays, if you put your money into a savings account or into an investment, you should be given the annual equivalent rate (AER). You may have invested your money for any period of time, but the AER tells you how much your money would earn in **exactly one year**. For example, suppose a bank offers you a five-year deal: 4% interest for the first six months followed by 2% for each of the remainder of the time. It will be much simpler to compare this deal with others if you can compare them on some common standard. The AER would, in this case, give you one simple annual percentage to use for comparison.

When borrowing money, there are often other costs involved, such as a set-up fee. Because these other costs can be significant, lenders are expected to tell the borrower the annual percentage rate (APR). This again allows a potential borrower to compare different loans and see which is more expensive.

Two of the most common loans are mortgages (a loan to purchase a home) and credit cards.

Note: An APR for a five-year loan is likely to be different to that for a three-year loan, even from the same lender. This can occur because there may be a set-up fee or an introductory reduced interest rate. A set-up fee will have little effect on the APR if it is spread over more years.

EXAMPLE 1

Calculate the value of an investment of €20,000 for 12 years at an annual equivalent rate (AER) of 3%, correct to the nearest cent.

Solution:
As $t > 3$, it is easier to use the formula.

Given: $P = 20{,}000$, $i = \dfrac{3}{100} = 0\cdot03$, $t = 12$.   Find F.

$$\begin{aligned}F &= P(1 + i)^t \\ &= 20{,}000\,(1\cdot03)^{12} \\ &= 28{,}515\cdot21774 \\ &= 28{,}515\cdot22 \quad \text{(correct to the nearest cent)}\end{aligned}$$

The value of the investment = €28,515·22.

EXAMPLE 2

An investment of €50,000 will earn 15% interest after six years. Calculate the annual equivalent rate (AER), correct to two decimal places.

Solution:
The final amount = €50,000 × 1·15 = €57,500.
Given: $P = 50{,}000$, $F = 57{,}500$, $t = 6$.   Find i.
Rewrite the formula so that the unknown is on the left.

$$P(1 + i)^t = F \quad \text{(rewrite formula)}$$

$$50{,}000(1 + i)^6 = 57{,}500$$

$$(1 + i)^6 = \dfrac{57{,}500}{50{,}000} \quad \text{(divide both sides by 50,000)}$$

$$(1 + i) = \sqrt[6]{\dfrac{57{,}500}{50{,}000}} \quad \text{(get the sixth root of both sides)}$$

$$= 1\cdot023567073$$

$$= 1\cdot02 \quad \text{(correct to two decimal places)}$$

The annual equivalent rate (AER) = 1·02%.

Depending on the model of your calculator and noting some functions need [SHIFT] or [2nd F], one of the following might work. Check the manual for your calculator.

(older) 🖩 6 [$\sqrt[x]{}$] [(] 57,500 [÷] 50,000 [)] [=]

(newer) 🖩 [$\sqrt[x]{}$] 6 [RIGHT] [FRACTION] 57,500 [DOWN] 50,000 [RIGHT] [=]

Exercise 2.10

In questions 1–4, calculate the value of the investment to the nearest cent after the given period.

1. €20,000 for 8 years with an AER of 5%
2. €15,000 for 4 years with an AER of 7%
3. €36,000 for 6 years with an AER of 4%
4. €24,000 for 10 years with an AER of 2%

In questions 5–8, calculate the amount owing to the nearest cent after the given period.

5. €5,000 for 3 years with an APR of 6%
6. €12,000 for 5 years with an APR of 5%
7. €6,000 for 6 years with an APR of 4·5%
8. €4,000 for 4 years with an APR of $2\frac{1}{2}$%

9. (i) Calculate the interest gained by investing €200 for one year:
   (a) With 10% added at the end of the year
   (b) With 5% added after each six months
   (ii) What is the AER of each?
   (iii) How would the interest compare if the investment had $2\frac{1}{2}$% added every three months?

10. Calculate the value of investing €100 for each of the following.
    (i) One year with 10% interest paid at the end of the year
    (ii) One year with 6% interest paid after each six months
    (iii) 12 months, with 2% paid after every two months
    (iv) Deduce the annual equivalent rate (AER) for each.

11. (i) A credit card charges $1\frac{1}{2}$% interest on unpaid monthly balances. If €1,000 is borrowed at the beginning of a month and not paid for a year, how much will be owed?
    (ii) What should the credit card company declare as the APR for this card? Give your answer correct to one place of decimals.

12. (i) If $F = P(1 + i)^t$, express P in terms of F, i and t.
    (ii) Hence or otherwise, find what sum of money will amount to €88,578·05 in six years with an AER of 10%.

13. What sum of money will amount to €243,101·25 in four years with an AER of 5%?

14. An investment of €20,000 will earn you 25% interest over five years.
    (i) Calculate the annual equivalent rate (AER), correct to three decimal places.
    (ii) How much interest will you earn in the first two years? Answer to the nearest euro.
    (iii) Why is 25% over five years not the same as 5% annually?

15. Two people invest money for eight years with a return of 20% when the investment matures. Shane invests €200, while Tommy invests €300. Show that the annual equivalent rate for each investor is the same.

16. An investor is offered a return of 16% after seven years. Calculate the AER.
    (Hint: Use a simple amount as the investment: €100 or €1,000, for example.)

Distance, speed and time

There are three formulas to remember when dealing with problems involving distance (*D*), speed (*S*) and time (*T*). It can be difficult to remember these formulas; however, the work can be made easier using a triangle and the memory aid Dad's Silly Triangle.

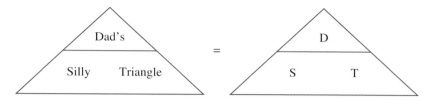

1. Speed = $\dfrac{\text{Distance}}{\text{Time}}$ 2. Time = $\dfrac{\text{Distance}}{\text{Speed}}$ 3. Distance = Speed × Time

Consider the triangle on the right. By covering the quantity required, *D*, *S* or *T*, any of the three formulas above can be found by inspection.

Note: Speed here means average speed.

Common units of speed

1. Kilometres per hour, written as km/h.
2. Metres per second, written as m/s.

Note: 'Per' means divided by.

Converting fractions or decimals of an hour to minutes

To convert fractions or decimals of an hour to minutes, **multiply by 60**.
For example:

$\frac{1}{3}$ hour = $\frac{1}{3}$ × 60 minutes = 20 minutes   0.7 hour = 0·7 × 60 minutes = 42 minutes

Converting minutes to hours

To convert minutes to hours, **divide by 60**.
For example:

15 minutes = $\frac{15}{60}$ hour = $\frac{1}{4}$ hour or 0·25 hour

48 minutes = $\frac{48}{60}$ hour = $\frac{4}{5}$ hour or 0·8 hour

EXAMPLE

(i) A train takes 3 hours and 30 minutes to travel 280 km. Calculate the average speed in km/h.

(ii) How long, in hours and minutes, does it take a bus to travel 168 km at an average speed of 96 km/h?

(iii) A car travelled at an average speed of 120 km/h between 12:55 and 14:10. What distance did it travel?

Solution:

(i) Time has to be expressed in hours.
3 hours and 30 minutes = $3\frac{1}{2}$ hours

$$\text{Speed} = \frac{\text{Distance}}{\text{Time}}$$

$$= \frac{280}{3\frac{1}{2}}$$

$$= 80 \text{ km/h}$$

(iii) 14 hrs 10 mins − 12 hrs 55 mins
= 1 hr 15 mins
= $1\frac{1}{4}$ hrs

(ii) $\text{Time} = \frac{\text{Distance}}{\text{Speed}}$

$$= \frac{168}{96}$$

= 1·75 hours
= 1 hour and 45 minutes
(0·75 hour = 45 minutes)

Distance = speed × time
= 120 × $1\frac{1}{4}$
= 150 km

Two-part problems

Two-part questions on distance, speed and time involve two separate journeys. In these questions we need the total distance travelled for both journeys and the total time for both journeys. We then use the formula:

$$\text{Overall average speed for both journeys} = \frac{\text{Total distance for both journeys}}{\text{Total time for both journeys}}$$

EXAMPLE

A train travelled 168 km at an average speed of 112 km/h. It then travelled for 45 minutes at an average speed of 100 km/h. Calculate:

(i) The total distance travelled
(ii) The total time taken
(iii) The average speed for the whole journey

Solution:

In the first journey, the time is required.

$$\text{Time} = \frac{\text{Distance}}{\text{Speed}}$$
$$= \frac{168}{112}$$
$$= 1\frac{1}{2} \text{ hours}$$

In the second journey, the distance is required.

$$\text{Distance} = \text{Speed by time}$$
$$= 100 \times \frac{3}{4}$$
$$= 75 \text{ km}$$

(Note: 45 minutes = $\frac{3}{4}$ hour)

(i) Total distance travelled = 168 + 75 = 243 km

(ii) Total time = $1\frac{1}{2} + \frac{3}{4} = 2\frac{1}{4}$ hours

(iii) Average speed for the whole journey = $\frac{\text{Total distance travelled}}{\text{Total time taken}}$
$= \frac{243}{2\frac{1}{4}} = 108$ km/h

Exercise 2.11

1. Express each of the following in minutes.
   (i) 0·5 hr   (ii) 0·6 hr   (iii) 0·1 hr   (iv) 0·7 hr   (v) 0·4 hr

2. Express each of the following in hours and minutes.
   (i) 1·2 hrs   (ii) $2\frac{3}{4}$ hrs   (iii) 3·35 hrs   (iv) $1\frac{1}{3}$ hrs   (v) 2·45 hrs

3. Express each of the following in hours.
   (i) 15 mins   (ii) 20 mins   (iii) 1 hr 30 mins   (iv) 2 hrs 45 mins   (v) 3 hrs 24 mins

4. A train takes 3 hours and 30 minutes to travel 280 km. Calculate the average speed in km/h.

5. How long, in hours and minutes, does it take a bus to travel 168 km at an average speed of 96 km/h?

6. A car travelled at an average speed of 120 km/h between 12:55 hrs and 14:10 hrs. What distance did it travel?

7. The distance, by rail, between Galway and Dublin is 240 km. On Tuesday, a train left Galway at 13:05 and travelled to Dublin. The average speed for this journey of 240 km was 100 km/h. At what time did the train arrive in Dublin?

8. Cormac went by car from Limerick to Cork, a journey of 100 km. He travelled at an average speed of 80 km/h.
   (i) How many hours and minutes did it take Cormac to complete the journey?
   (ii) Cormac left Limerick at 11:15. At what time did he arrive in Cork?
   (iii) Cormac's car used 1 litre of petrol for every 16 km travelled. On that day petrol cost €1·24 per litre. Find the cost of the petrol used on Cormac's journey from Limerick to Cork.

9. Find the average speed of a car that completes a journey of 208 km in 3 hours 15 minutes.

10. A girl ran a distance of 3,600 m in 15 minutes. Calculate her average speed in m/s.

11. A motorcyclist begins a journey of 280 km at 15:00. If the average speed is 80 km/h, find the time at which the journey is completed.

12. (i) Express 4 hours 15 minutes in hours.
    (ii) A train starts a journey of 255 km at 09:40 and completes the journey at 13:55. What was the average speed of the train?

13. A car travels 54 km in 45 minutes. Calculate its average speed in km/h.

14. A train travels 52 km in 40 minutes. How far will it travel in 1 hour 10 minutes at the same average speed?

15. Anne walks a distance of 1·7 km to school from home. She walks at an average speed of 5·1 km/h. What is the latest time she can leave home to be in school at 08:55?

16. A train travels 180 m in 10 seconds. What is its speed in (i) m/s   (ii) km/h?

17. A car travels 28 km at an average speed of 80 km/h. How long does the journey take?

18. A cyclist travels for 45 minutes at an average speed of 19 km/h. What distance does the cyclist travel?

19. Mary leaves her home at 06:40. She drives 87 km to work and she travelled at an average speed of 36 km/h. At what time did she arrive at work?

20. In a athletics match, the 100 m race was won in a time of 9·92 s and the 200 m race was won in a time of 19·92 s. Which race was won with the slowest average speed? Justify your answer.

21. A cyclist and a bus leave the same place at the same time to the same destination. However, the cyclist and the bus take two different routes. The cyclist's route takes 15 minutes at an average speed of 20 km/h. The route for the bus is 3 km longer. Calculate the average speed of the bus if it is to arrive at the destination at the same time.

22. When a cyclist had travelled a distance of 12·6 km, he had completed $\frac{3}{7}$ of his journey. What was the length of the journey?

23. In a 4 × 400 m relay race, the split times of the runners for each leg of a team are shown in the table.
    Calculate, correct to two decimal places, in m/s:
    (i) The average speed of each runner
    (ii) The average speed of the relay team

First leg	46·15
Second leg	45·36
Third leg	45·04
Fourth leg	44·82

24. Frank drove from his home to his friend's house. He drove the first 3 hours at an average speed of 60 km/h. He then drove the remaining 90 km at an average speed of 45 km/h. Calculate Frank's average speed for the whole journey.

25. A car travels 100 km at an average speed of 50 km/h and then 180 km at an average speed of 60 km/h. Calculate: **(i)** The total distance travelled **(ii)** The total time taken
    (iii) The overall average speed

26. A bus travels for 300 km at an average speed of 75 km/h and then 162 km at an average speed of 81 km/h. Calculate: **(i)** The total time taken **(ii)** The total distance travelled
    (iii) The overall average speed

27. A woman drove her car for 200 km at an average speed of 80 km/h and then for 60 km at an average speed of 40 km/h. Find: **(i)** The total time taken **(ii)** The total distance travelled
    (iii) The overall average speed

28. On an outward journey of 280 km, a driver takes $3\frac{1}{2}$ hours. On the return journey she takes a short cut, driving at an average speed of 104 km/h, and this takes one hour less than the outward journey. Calculate the overall average speed.

29. A lorry travels 189 km in $3\frac{1}{2}$ hours. It then travels for $4\frac{1}{2}$ hours at an average speed of 70 km/h.
    (i) How far has it travelled altogether?
    (ii) What is the total time taken?
    (iii) What is the overall average speed?

30. On an outward journey of 330 km a driver takes 6 hours, and on the return journey he takes one hour less. Calculate:
    (i) The average speed of the outward journey
    (ii) The average speed of the return journey
    (iii) The total time and the total distance
    (iv) The overall average speed

Universal Social Charge (USC)

A new tax, the Universal Social Charge (USC), was introduced in 2011 to be applied to gross income. Initially, the rates were:

    0% if total annual income is €4,004 or under.

However, once over €4,004, the **full** annual income is taxed as:

    2% for the first €10,036

    4% for the next €5,980

    7% on the balance.

The above information could also be expressed in terms of thresholds or cut-off points:

    2% if salary is over €4,004

    4% if salary is over €10,036

    7% if salary is over €16,016

EXAMPLE

A woman earns €36,000 and is required to pay the Universal Social Charge (USC) as follows:

    2% for the first €10,036

    4% for the next €5,980

    7% on the balance

What amount must she pay?

Solution:

Her income is over €10,036, so €36,000 − €10,036 = €25,964 must be paid at a higher rate. This is more than €5,980, so €25,964 − €5,980 = €19,984 must be paid at the highest rate. Her income must be split into three separate amounts:

Income (€)	Rate	Calculation	USC (€)
10,036	2%	10,036 × 0·02	200·72
5,980	4%	5,980 × 0·04	239·20
19,984	7%	19,984 × 0·07	1,398·88
36,000			1,838·80

She must pay €1,838·80.

Exercise 2.12

In questions 1–8, calculate the Universal Social Charge on the given salary if the rates are:

> 0% if under €5,000
> 2% for the first €10,000 of the full salary
> 3% for the next €6,000
> 6% on the balance

1. €8,000
2. €15,000
3. €25,000
4. €12,000
5. €20,000
6. €4,000
7. €160,000
8. €100,000

In questions 9–16, calculate the Universal Social Charge on the given salary if the rates are:

> 0% if €4,004 or under
> 2% for the first €10,036 of the full salary
> 4% for the next €5,980
> 7% on the balance

9. €9,500
10. €7,500
11. €14,000
12. €28,000
13. €40,000
14. €90,000
15. €240,000
16. €4,010

17. Two students work during their summer break, earning €100 per weekday and €150 on Saturdays. Alex works for eight weeks but does not work on Saturdays. Bruce works for seven weeks including Saturdays.

    (i) What is the gross earnings of each student?

    (ii) A Universal Social Charge (0% if €4,004 or under but 2% for the first €10,036 of the full salary if over €4,004) is applied. What is the Universal Social Charge for each student?

    (iii) Alex had considered working for one extra day on the Monday of the ninth week. Why might she have decided not to work?

    (iv) How much extra would she have earned for that day's work?

Income tax

The following is called the income tax equation:

$$\text{Gross tax} - \text{tax credit} = \text{tax payable}$$

Gross tax is calculated as follows:

$$\boxed{\text{Standard rate on all income up to the standard rate cut-off point}} + \boxed{\text{A higher rate on all income above the standard rate cut-off point}}$$

EXAMPLE 1

A woman has a gross yearly income of €48,000. She has a standard rate cut-off point of €27,500 and a tax credit of €3,852. The standard rate of tax is 18% of income up to the standard rate cut-off point and 37% on all income above the standard rate cut-off point. Calculate:

(i) The amount of gross tax for the year
(ii) The amount of tax paid for the year

Solution:

(i) Gross tax = 18% of €27,500 + 37% of €20,500
     = €27,500 × 0.18 × €20,500 + 0·37
     = €4,950 + €7,585
     = €12,535

Income above the standard rate cut-off point
= €48,000 − €27,500
= €20,500

(ii) Income tax equation:

Gross tax − tax credit = tax payable
€12,535 − €3,852 = €8,683

Therefore, she paid €8,683 in tax.

Note: If a person earns less than their standard rate cut-off point, then they pay tax only at the standard rate on all their income.

EXAMPLE 2

A man paid €10,160 in tax for the year. He had a tax credit of €3,980 and a standard rate cut-off point of €26,000. The standard rate of tax is 17% of income up to the standard rate cut-off point and 36% on all income above the standard rate cut-off point. Calculate:

(i) The amount of income taxed at the rate of 36%

(ii) The man's gross income for the year

Solution:

(i) Income tax equation:

$$\text{Gross tax} - \text{tax credit} = \text{tax payable}$$
$$17\% \text{ of } €26{,}000 + 36\% \text{ of (income above cut-off point)} - €3{,}980 = €10{,}160$$
$$€4{,}420 + 36\% \text{ of (income above cut-off point)} - €3{,}980 = €10{,}160$$
$$36\% \text{ of (income above cut-off point)} + €440 = €10{,}160$$
$$36\% \text{ of (income above cut-off point)} = €9{,}720$$
$$1\% \text{ of (income above cut-off point)} = €270$$
(divide both sides by 36)
$$100\% \text{ of (income above cut-off point)} = €27{,}000$$
(multiply both sides by 100)

Therefore, the amount of income taxed at the higher rate of 36% was €27,000.

(ii) Gross income = standard rate cut-off point + income above the standard rate cut-off point
= €26,000 + €27,000 = €53,000

Exercise 2.13

In questions 1–4, make out a tax table similar to the following:

Gross pay	
Tax @ %	
Tax @ %	
Gross tax	
Tax credit	
Tax payable	
Take-home pay	

1. A woman has a gross yearly income of €39,000. She has a standard rate cut-off point of €24,000 and a tax credit of €3,800. The standard rate of tax is 20% of income up to the standard rate cut-off point and 42% on all income above the standard rate cut-off point. Calculate:
   (i) The amount of gross tax for the year
   (ii) The amount of tax paid for the year

2. A man has a gross yearly income of €37,000. He has a standard rate cut-off point of €20,500 and a tax credit of €2,490. The standard rate of tax is 18% of income up to the standard rate cut-off point and 38% on all income above the standard rate cut-off point. Calculate:
   (i) The amount of gross tax for the year
   (ii) The amount of tax paid for the year

3. A man has a gross yearly income of €43,000. He has a standard rate cut-off point of €28,400 and a tax credit of €3,240. The standard rate of tax is 15% of income up to the standard rate cut-off point and 35% on all income above the standard rate cut-off point. Calculate:
   (i) The amount of gross tax for the year
   (ii) The amount of tax paid for the year

4. A woman has a gross yearly income of €48,700. She has a standard rate cut-off point of €29,250 and a tax credit of €3,150. The standard rate of tax is 16% of income up to the standard rate cut-off point and 37% on all income above the standard rate cut-off point. Calculate:
   (i) The amount of gross tax for the year
   (ii) The amount of tax paid for the year

5. A man has a gross yearly income of €26,000. He has a standard rate cut-off point of €28,000 and a tax credit of €1,800. If he pays tax of €3,400, calculate the standard rate of tax.

6. A woman has a gross yearly income of €27,500. She has a standard rate cut-off point of €29,300 and a tax credit of €2,115. If she pays tax of €2,835, calculate the standard rate of tax.

7. A woman paid €10,280 in tax for the year. She had a tax credit of €2,540 and a standard rate cut-off point of €29,000. The standard rate of tax is 18% of income up to the standard rate cut-off point and 40% on all income above the standard rate cut-off point. Calculate:
   (i) The amount of income taxed at the rate of 40%
   (ii) The gross income for the year

8. A man paid €10,775 in tax for the year. He had a tax credit of €1,960 and a standard rate cut-off point of €28,500. The standard rate of tax is 15% of income up to the standard rate cut-off point and 36% on all income above the standard rate cut-off point. Calculate:

   (i) The amount of income taxed at the rate of 36%

   (ii) The gross income for the year

The USC and income tax

Both taxes are applied to the **gross salary** and could be applied in either. We will calculate the USC first followed by the income tax payable.

EXAMPLE

Enda earns €200,000 in basic pay. The Universal Social Charge (USC) is as follows:

   2% for the first €10,036

   4% for the next €5,980

   7% on the balance

The standard rate of tax is 20% of income up to the standard rate cut-off point of €35,000 and 41% on all income above the standard rate cut-off point. Tax credits amount to €4,500. Calculate Enda's take-home pay.

Solution:
Calculate the USC as before:

Income (€)	Rate	Calculation	USC (€)
10,036	2%	10,036 × 0·02	200·72
5,980	4%	5,980 × 0·04	239·20
183,984	7%	183,984 × 0·07	12,878·88
200,000			13,318·80

Then calculate the income tax and include the USC.

Gross pay		200,000·00	
Tax @ 20% on €35,000	7,000		
Tax @ 41% on €165,000	67,650		
Gross tax	74,650		(add the two tax amounts)
Tax credit	4,500		
Tax payable		70,150·00	(subtract the tax credit)
USC		13,318·80	(from above)
Total payable		83,468·80	(add the two taxes)
Take-home pay		116,531·20	(subtract taxes from gross pay)

Exercise 2.14

In questions 1–4, calculate the take-home pay for the given gross annual salary and tax credit. The Universal Social Charge is calculated as follows:

> 0% if under €5,000
> 2% for the first €10,000 of the full salary
> 3% for the next €6,000
> 6% on the balance

The standard rate of tax is 20% of income up to the standard rate cut-off point of €32,000 and 40% on all income above the standard rate cut-off point.

1. Salary: €10,000; tax credit: €1,500
2. Salary: €45,000; tax credit: €4,200
3. Salary: €80,000; tax credit: €5,200
4. Salary: €120,000; tax credit: €4,800

In questions 5–8, calculate the take-home pay for the given gross annual salary. The Universal Social Charge is calculated as follows:

> 0% if €4,004 or under
> 2% for the first €10,036 of the full salary
> 4% for the next €5,980
> 7% on the balance

The standard rate of tax is 21% of income up to the standard rate cut-off point of €28,000 and 41% on all income above the standard rate cut-off point.

5. Salary: €15,000; tax credit: €3,600
6. Salary: €38,000; tax credit: €5,000
7. Salary: €150,000; tax credit: €6,100
8. Salary: €300,000; tax credit: €7,500

Index notation

Index notation is a shorthand way of writing very large or very small numbers. For example, try this multiplication on your calculator: $8,000,000 \times 7,000,000$.

The answer is $56,000,000,000,000$.
It has 14 digits, which is too many to show on most calculator displays.

Your calculator will display your answer as $\boxed{5 \cdot 6 \text{ E } 13}$ or $\boxed{5 \cdot 6 \times 10^{13}}$ or $\boxed{5 \cdot 6^{13}}$
This tells you that the $5 \cdot 6$ is multiplied by 10^{13}.
This is written as:

$$5 \cdot 6 \times 10^{13}$$

This part is a number between 1 and 10 (but not including 10).

This part is written as a power of 10 (the power is always a whole number).

Another example to try on your calculator is: $0 \cdot 000\ 000\ 23 \times 0 \cdot 000\ 000\ 04$.
The answer is $0 \cdot 000\ 000\ 000\ 000\ 009\ 2$.

Your calculator will display your answer as $\boxed{9 \cdot 2 \text{ E} -15}$ or $\boxed{9 \cdot 2 \times 10^{-15}}$ or $\boxed{9 \cdot 2^{-15}}$

This tells you that the $9 \cdot 2$ is multiplied by 10^{-15}.
This is written as:

$$9 \cdot 2 \times 10^{-15}$$

This way of writing a number is called **index notation** or **exponential notation**, or sometimes **standard form**. (It was formerly called **scientific notation**.)

Index notation gives a number in two parts

$$\boxed{\text{Number between 1 and 10 (but not 10)}} \times \boxed{\text{Power of 10}}$$

This is often written as $a \times 10^n$, where $1 \leq a < 10$ and $n \in \mathbb{Z}$.

EXAMPLE 1

Express 3,700,000 in the form $a \times 10^n$, where $1 \leq a < 10$, $n \in \mathbb{Z}$.

Solution:

3,700,000· (put in the decimal point)

3·700 000 (move the decimal point six places to the **left**
to give a number between 1 and 10)

$\therefore 3{,}700{,}000 = 3 \cdot 7 \times 10^6$

EXAMPLE 2

Express the number 0·000846 in the form $a \times 10^n$, where $1 \leq a < 10$, $n \in \mathbb{Z}$.

Solution:

0·000846 (decimal point already there)

8·46 (move the decimal point four places to the **right**
to give a number between 1 and 10)

$\therefore 0 \cdot 000846 = 8 \cdot 46 \times 10^{-4}$

EXAMPLE 3

(i) Express $\dfrac{1{,}456}{0.28}$ in the form $a \times 10^n$, where $1 \leq a < 10$, $n \in \mathbb{Z}$.

(ii) Find n if $\dfrac{441}{0 \cdot 007} = 6 \cdot 3 \times 10^n$.

Solution:

(i) $\dfrac{1{,}456}{0 \cdot 28} = 5{,}200 = 5 \cdot 2 \times 10^3$

(ii) $\dfrac{441}{0 \cdot 007} = 63{,}000 = 6 \cdot 3 \times 10^4$

By comparing $6 \cdot 3 \times 10^n$ to $6 \cdot 3 \times 10^4$,

$n = 4$.

Exercise 2.15

In questions 1–16, evaluate and express the result in the form $a \times 10^n$, where $1 \leq a < 10$ and $n \in \mathbb{Z}$.

1. 8,000
2. 54,000
3. 347,000
4. 470
5. 2,900
6. 3,400,000
7. 394
8. 39
9. 0·006
10. 0·0009
11. 0·052
12. 0·000432
13. $\dfrac{1{,}512}{0{\cdot}36}$
14. $\dfrac{624}{0{\cdot}008}$
15. $\dfrac{0{\cdot}0048}{0{\cdot}15}$
16. $\dfrac{0{\cdot}0099}{2{\cdot}2}$

In questions 17–19, calculate the value of n.

17. $\dfrac{2{,}856}{0{\cdot}42} = 6{\cdot}8 \times 10^n$
18. $\dfrac{73{,}080}{1{\cdot}74} = 4{\cdot}2 \times 10^n$
19. $\dfrac{0{\cdot}0624}{2{\cdot}6} = 2{\cdot}4 \times 10^n$

Using a calculator

Most scientific calculators can be set to **display all answers** in index (scientific) notation. The procedure varies with different models and different manufacturers, so you are advised to read your calculator's manual. Furthermore, you will need to be able to return your calculator to its normal display settings.

Calculators that have a $\boxed{\text{SET UP}}$ button (which may need to be preceded with $\boxed{\text{SHIFT}}$) may offer you either an option **FSE** or take you to a list of display options. Selecting the **FSE** option may also take you to a list of display options.

The usual display options include **FIX**ed decimal place, **SCI**entific notation and **NORM**al. Using the **SCI** option will cause all answers to be displayed in index notation and the calculator screen should show **SCI** to confirm the display mode. You may continue to enter numbers in the usual manner.

To return your display to its usual state, you will need to need to go through the procedure again, this time choosing **NORM**al display. Most calculators have two versions of **NORM**al, so you may have to select **1** or **2**. The calculator screen will no longer show the **SCI** indicator.

Calculators that do not have a $\boxed{\text{SET UP}}$ button should have a MODE button, which if pressed repeatedly will provide display options.

Notes:
1. The display modes only refer to how the **answer** is displayed. You may enter numbers in any format at all times.
2. Very large and very small numbers are always displayed in index notation.
3. Remember to set your calculator back to **NORM**al display mode.

Addition and subtraction

Numbers given in index notation can be keyed into your calculator by using the **exponent key**. It is marked $\boxed{\text{EXP}}$ or $\boxed{\text{EE}}$ or $\boxed{\times 10^x}$.

To key in a number in index notation, do the following:

> 1. Key in 'a', the 'number part', first.
> 2. Press the exponent key next.
> 3. Key in the index of the power of 10.

To enter $3 \cdot 4 \times 10^6$, for example, you key in $3 \cdot 4$ $\boxed{\text{EXP}}$ 6.

To enter negative powers, you need to find the **negative** button on your calculator. It is usually marked $\boxed{(-)}$ or $\boxed{+/-}$ and is used to enter negative numbers.

To enter $7 \cdot 1 \times 10^{-3}$, for example, you key in $7 \cdot 1$ $\boxed{\text{EXP}}$ $\boxed{(-)}$ 3.

Note: If you press $\boxed{=}$ at the end, the calculator will write the number as a decimal number, provided the index of the power of 10 is not too large.

To add or subtract two numbers in index notation, do the following:

> 1. Write each number as a simple number.
> 2. Add or subtract these numbers.
> 3. Write your answer in index notation.
>
> Alternatively, you can use your calculator by keying in the numbers in index notation and adding or subtracting as required.

EXAMPLE 1

Express $2 \cdot 54 \times 10^4 - 3 \cdot 8 \times 10^3$ in the form $a \times 10^n$, where $1 \leq a < 10$ and $n \in \mathbb{Z}$.

Solution:

$$2 \cdot 54 \times 10^4 = 25{,}400$$
$$3 \cdot 8 \times 10^3 = \underline{3{,}800}$$
$$= 21{,}600 \quad \text{(subtract)}$$
$$= 2 \cdot 16 \times 10^4$$

🖩 $2 \cdot 54$ $\boxed{\text{EXP}}$ 4 $\boxed{-}$ $3 \cdot 8$ $\boxed{\text{EXP}}$ 3 $\boxed{=}$ $= 21{,}600$ (on the display) $= 2 \cdot 16 \times 10^4$

EXAMPLE 2

Express $2.68 \times 10^2 + 1.2 \times 10^3$ in the form $a \times 10^n$, where $1 \leq a < 10$ and $n \in \mathbb{Z}$.

Solution:

$$2.68 \times 10^{-2} = 0.0268$$
$$1.2 \times 10^{-3} = \underline{0.0012}$$
$$= 0.0280 \quad \text{(add)}$$
$$= 2.8 \times 10^{-2}$$

🖩 2.68 [EXP] [(−)] 2 [+] 1.2 [EXP] [(−)] 3 [=] = 0.028 (on the display) = 2.8×10^{-2}

Multiplication and division

To multiply or divide two numbers in index notation, do the following:

1. Multiply or divide the 'a' parts (the number parts).
2. Multiply or divide the powers of 10 (add or subtract the indices).
3. Write your answer in index notation.

Alternatively, you can use your calculator by keying in the numbers in index notation and multiplying or dividing as required.

EXAMPLE

Express: **(i)** $(3.5 \times 10^2) \times (4.8 \times 10^3)$ **(ii)** $(4.86 \times 10^4) - (1.8 \times 10^7)$ in the form $a \times 10^n$, where $1 \leq a < 10$ and $n \in \mathbb{Z}$.

Solution:

(i)
$$(3.5 \times 10^2) \times (4.8 \times 10^3)$$
$$= 3.5 \times 10^2 \times 4.8 \times 10^3$$
$$= 3.5 \times 4.8 \times 10^2 \times 10^3$$
$$= 16.8 \times 10^{2+3} \quad \text{(add the indices)}$$
$$= 16.8 \times 10^5$$
$$= 1{,}680{,}000$$
$$= 1.68 \times 10^6$$

🖩 3.5 [EXP] 2 [×] 4.8 [EXP] 3 [=]
= 1,680,000 (on the display)
= 1.68×10^6

(ii)
$$(4.86 \times 10^4) \div (1.8 \times 10^7)$$
$$= \frac{4.86 \times 10^4}{1.8 \times 10^7}$$
$$= \frac{4.86}{1.8} \times \frac{10^4}{10^7}$$
$$= 2.7 \times 10^{4-7} \quad \text{(subtract the indices)}$$
$$= 2.7 \times 10^{-3}$$

🖩 4.86 [EXP] 2 [÷] 1.8 [EXP] 7 [=]
= 0.0027 (on the display)
= 2.7×10^3

Exercise 2.16

In questions 1–22, simplify and express your answer in the form $a \times 10^n$, where $1 \leq a < 10$ and $n \in \mathbb{Z}$.

1. $2 \cdot 4 \times 10^3 + 8 \times 10^2$
2. $2 \cdot 52 \times 10^6 + 2 \cdot 8 \times 10^5$
3. $5 \cdot 48 \times 10^5 - 2 \cdot 8 \times 10^4$
4. $48 \cdot 2 \times 10^3 - 2 \cdot 52 \times 10^4$
5. $8 \cdot 45 \times 10^{-3} - 6 \cdot 5 \times 10^{-4}$
6. $3 \cdot 48 \times 10^{-4} - 5 \cdot 4 \times 10^{-5}$
7. $(1 \cdot 8 \times 10^3) \times (4 \times 10^4)$
8. $(2 \cdot 25 \times 10^4) \times (1 \cdot 6 \times 10^3)$
9. $(2 \cdot 2 \times 10^3) \times (3 \cdot 4 \times 10^2)$
10. $(5 \cdot 3 \times 10^2) \times (1 \cdot 8 \times 10^4)$
11. $(3 \cdot 91 \times 10^5) \div (1 \cdot 7 \times 10^2)$
12. $(5 \cdot 04 \times 10^7) \div (3 \cdot 6 \times 10^2)$
13. $(8 \cdot 64 \times 10^5) \div (3 \cdot 6 \times 10^2)$
14. $(9 \cdot 86 \times 10^5) \div (1 \cdot 7 \times 10^2)$
15. $(3 \times 10^3) \div (2 \times 10^{-2})$
16. $(12 \cdot 6 \times 10^3) \div (4 \cdot 5 \times 10^7)$
17. $(5 \cdot 4 \times 10^2) \times (6 \cdot 5 \times 10^3)$
18. $(1 \cdot 35 \times 10^7) \div (2 \cdot 5 \times 10^3)$
19. $\dfrac{(2 \cdot 4 \times 10^4) \times (1 \cdot 5 \times 10^2)}{1 \cdot 2 \times 10^3}$
20. $\dfrac{(3 \cdot 2 \times 10^5) + (8 \cdot 5 \times 10^4)}{8 \cdot 1 \times 10^2}$
21. $\dfrac{2 \cdot 45 \times 10^5 - 1 \cdot 8 \times 10^3}{1 \cdot 6 \times 10^3}$
22. $\dfrac{1 \cdot 4 \times 10^3 + 5 \cdot 6 \times 10^2}{7 \times 10^{-1}}$

23. Calculate the value of $8 \cdot 45 \times 10^{-2} - 6 \cdot 5 \times 10^{-3}$.
    Write your answer as a decimal number.
    Say whether this number is greater than or less than $0 \cdot 08$.

24. Calculate the value of $\dfrac{2 \cdot 8 \times 10^4 + 4 \cdot 2 \times 10^5}{2 \cdot 24 \times 10^6}$.
    Write your answer as a decimal number.
    Say whether this number is greater than or less than $0 \cdot 19$.

25. $\sqrt{\dfrac{3 \cdot 64 \times 10^5 - 1 \cdot 7 \times 10^3}{9 \cdot 0575 \times 10^2}} = k$. Find the value of k.

Miscellaneous

Exercise 2.17

1. The owner of a small business agrees to give his staff an annual pay increase of 4% for the next three years.
   (i) If Paul is currently earning €23,000, what will his salary be (to the nearest euro) in three years' time?

(ii) Peter began employment just over 10 years ago. He began with a salary of €14,000 with a 3% increase per year. What is Peter's current salary to the nearest euro?

2. The population of a colony of rabbits in 2010 was 450. Allowing for births and deaths, it is estimated that the colony will grow by 12% annually. What is the expected population by 2020, to the nearest integer?

3. I invest €100,000 at an APR of 6%.
   (i) What will my investment be worth after 30 years (to the nearest cent)?
   (ii) What will my investment be worth after 35 years (to the nearest cent)?
   (iii) How long will it take to become €1 million?

4. Show that an APR of 10% is equivalent to 21% over two years.

5. A gift of €100 is invested for 21 years at an AER of 3%. What will it be worth (to the nearest euro) when it matures?

6. Harry invests €1,000 in a savings account with an AER of 4%.
   (i) Harry thinks it will double in value in 25 years. Explain why he is wrong.
   (ii) How much will it be worth in 25 years' time?
   (iii) At what rate would Harry's investment double in 25 years? Give your answer correct to one decimal place.

7. Jim is paid €7·50 an hour. The table shows the hours he worked during one week.

Day	Start	Finish
Monday	09:30	12:30
Tuesday	14:00	17:30
Friday	17:00	22:00

   (i) How many hours did Jim work?
   (ii) How much did he earn?

8. Holiday-makers who book with Sunny Travel and who cancel a holiday have to pay a charge. The cancellation charge depends on the number of days before the departure date when the customer cancels the holiday.

   The cancellation charge is a percentage of the cost of the holiday, as shown in the table.

Number of days before departure date	Charge as a percentage of cost of holiday
29–55	40%
22–28	60%
15–21	80%
4–14	90%
3 or less	100%

   (i) Bertie's holiday cost €650. He cancelled his holiday 26 days before the departure date. Calculate Bertie's cancellation charge.
   (ii) Joan's holiday cost €1,200. She had to pay a charge of €960. Estimate when she cancelled her holiday.

(iii) Ronald cancelled his holiday 30 days before the departure date. He had to pay a cancellation charge of €504. Calculate the cost of his holiday.

9. Alison pays a fixed monthly charge of €16 for her mobile phone. This charge includes 200 free text messages and 50 minutes free call time each month. Further call time costs 28 cent per minute and additional text messages cost 11 cent each. In one month Alison sends 240 text messages and her call time is $2\frac{1}{2}$ hours.

   (i) Find the total cost of her fixed charge, text messages and call time.

   (ii) VAT is added to this cost at the rate of 21%. Find the amount paid, including VAT, to the nearest cent.

10. A supermarket has a special offer on three different brands of packets of soap. The following table gives details of the offer. Which brand has the cheapest price per gram?

Brand	No. of bars per packet	Weight of each bar	Price of packet
A	3	100g	€1·35
B	6	100g	€2·40
C	4	125g	€2·38

In questions 11–27, express your answer in the form $a \times 10^n$, where $1 \leq a < 10$ and $n \in \mathbb{Z}$, unless otherwise instructed.

11. The base of a microchip is in the shape of a rectangle. Its length is 2×10^{-3} mm and its width is $1·4 \times 10^{-3}$ mm. Find the area of the base of the microchip.

12. (i) Express 1·5 cm in m.

    (ii) Light travels at 3×10^8 m/s. Calculate the time it takes to travel 1·5 cm.

13. Given that $x = 2 \times 10^{-3}$ and $y = 7 \times 10^{-4}$, evaluate $x + 8y$.

14. (i) A floppy disk can store 1,440,000 bytes of data. Write the number 1,440,000 in the form $a \times 10^n$, where $1 \leq a < 10$ and $n \in \mathbb{Z}$.

    (ii) A hard disk can store $5·112 \times 10^9$ bytes of data. Calculate the number of floppy disks needed to store $4·8 \times 10^9$ bytes.

15. (i) Light travels at a speed of approximately 3×10^8 m/s. How many kilometres will light travel in 8 minutes?

    (ii) The Andromeda galaxy is 21,900,000,000,000,000,000 km from Earth.

    Write this distance in the form $a \times 10^n$, where $1 \leq a < 10$ and $n \in \mathbb{Z}$, correct to two significant figures.

(iii) Calculate the number of years light takes to travel from the Andromeda galaxy to Earth, using 1 year = 365·25 days. Express your answer in the form $a \times 10^n$, where $1 \le a < 10$ and $n \in \mathbb{Z}$, correct to two significant figures.

16. New York City produces more waste per person than any other city in the world. Each person produces an average of 1·6 kg of waste per day. The population of New York is about $1·5 \times 10^7$. Calculate, correct to two significant figures, the number of tonnes of waste produced in a year for New York City. (Note: 1 tonne = 10^3 kg. 1 year = 365 days.)

17. If $f = 5 \times 10^{-6}$, express $\frac{1}{f}$ in the form $a \times 10^n$, where $1 \le a < 10$ and $n \in \mathbb{N}$.

18. The population density of a country is the average number of people per square kilometre. Sudan has a population of $2·75 \times 10^7$ and an area of $2·5 \times 10^6$ square kilometres. Calculate the population density of Sudan.

19. (i) A packet of A4 paper contains 5×10^2 sheets of paper. The packet is 6 cm in height. Calculate the thickness of one sheet of paper.

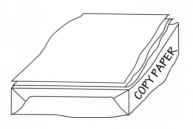

(ii) A magazine is made from 54 sheets of these A4 sheets of paper. The number of magazines printed is $6·48 \times 10^7$. Calculate the number of sheets of paper needed to print these magazines.

(iii) If all the magazines were piled up on top of each other, how high would the pile be? Give your answer to the nearest kilometre.

20. Express 2^{24} in the form $a \times 10^n$, where $1 \le a < 10$ and $n \in \mathbb{N}$, correct to three significant figures.

21. The mass of Jupiter is $1·91 \times 10^{27}$ kg and the mass of Earth is $5·97 \times 10^{24}$ kg. How many times greater is the mass of Jupiter than the mass of Earth? Give your answer to an appropriate level of accuracy.

22. The circumference of the Earth at the equator is about 4×10^4 km. Calculate the length of the radius at the equator. Express your answer in the form $a \times 10^n$, where $1 \le a < 10$ and $n \in \mathbb{Z}$, correct to three significant figures.

23. In 1981 the population of Peru was approximately $1·8 \times 10^7$. By 1988 the population had increased by 2·5 million. What would be the approximate population of Peru in 1988?

24. A postage stamp weighs $3·2 \times 10^{-5}$ kg. A speck of dust weighs $1·6 \times 10^{-7}$ kg. How many specks of dust weigh the same as a stamp?

25. The weight of an oxygen atom is $2 \cdot 7 \times 10^{-23}$ g and the weight of an electron is 9×10^{-28} g. If $k = \dfrac{\text{weight of an oxygen atom}}{\text{weight of an electron}}$, calculate the value of k.

26. The surface area of the Earth is approximately $5 \cdot 2 \times 10^{14}$ m^2. Approximately 30% of the surface area is land. What is the approximate area of the Earth that is covered by water?

27. Calculate h, the length of the hypotenuse of the right-angled triangle shown.

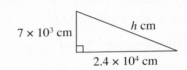

CHAPTER 3: COMPLEX NUMBERS

Imaginary numbers, the symbol i

Consider the equations (i) $x^2 - 1 = 0$ and (ii) $x^2 + 1 = 0$.

(i) $x^2 - 1 = 0$
$x^2 = 1$
$x = \pm\sqrt{1}$
$x = \pm 1$

(ii) $x^2 + 1 = 0$
$x^2 = -1$
$x = \pm\sqrt{-1}$

The solution to the second equation, $x^2 + 1 = 0$, requires finding $\sqrt{-1}$. This is the problem. To overcome this problem, mathematicians invented a new number. They defined:

$$i = \sqrt{-1} \quad \text{or} \quad i^2 = -1$$

$x^2 + 1 = 0$
$i^2 + 1 = 0$ (replace x with i)
$-1 + 1 = 0$ (true, where $i^2 = -1$)

The square root of a negative number is called an **imaginary** number, e.g. $\sqrt{-4}, \sqrt{-9}, \sqrt{-64}, \sqrt{-100}$ are imaginary numbers.

Imaginary numbers cannot be represented by a real number, as there is no real number whose square is a negative number.

All imaginary numbers can now be expressed in terms of i, for example:

$\sqrt{-36} = \sqrt{36 \times -1} = \sqrt{36}\sqrt{-1} = 6i$
$\sqrt{-81} = \sqrt{81 \times -1} = \sqrt{81}\sqrt{-1} = 9i$

Exercise 3.1

Express each of the following in the form ki, where $k \in \mathbb{N}$.

1. $\sqrt{-16}$
2. $\sqrt{-9}$
3. $\sqrt{-4}$
4. $\sqrt{-25}$
5. $\sqrt{-64}$
6. $\sqrt{-100}$
7. $\sqrt{-49}$
8. $\sqrt{-144}$

Complex numbers

A complex number has two parts, a **real** part and an **imaginary** part.
Some examples are $3 + 4i$, $2 - 5i$, $-6 + 0i$, $0 - i$.
Consider the complex number $4 + 3i$:

  4 is called the **real** part.
  3 is called the **imaginary** part.

Note: $3i$ is **not** the imaginary part.

> Complex number = (real part) + (imaginary part) i

The set of complex numbers is denoted by $\mathbb{C}$.
The letter z is usually used to represent a complex number. For example:

$$z_1 = 2 + 3i \quad z_2 = -2 - i \quad z_3 = -5i$$

If $z = a + bi$, then:
 (i) a is called the real part of z and is written $Re(z) = a$.
 (ii) b is called the imaginary part of z and is written $Im(z) = b$.

Note: u, v and w are also often used to denote complex numbers.

EXAMPLE

Write down the real and imaginary parts of each of the following complex numbers.
  (i) $3 + 2i$ (ii) $-6 - 8i$ (iii) 7 (iv) $-5i$

Solution:

	Real part	Imaginary part
(i) $3 + 2i$	3	2
(ii) $-6 - 8i$	-6	-8
(iii) $7 = 7 + 0i$	7	0
(iv) $-5i = 0 - 5i$	0	-5

Note: i **never** appears in the imaginary part.

Exercise 3.2
Write down the real and imaginary parts of each of the following complex numbers.

1. $5 + 3i$
2. $2 + 5i$
3. $6 + 7i$
4. $5 + 4i$
5. $2 - 7i$
6. $-4 + 6i$
7. $-3 - 5i$
8. $-9 + 8i$
9. $2 + i$
10. $3 - i$
11. $-5 + i$
12. $-1 - i$
13. 6
14. $2i$
15. -2
16. $-5i$

Addition, subtraction and multiplication by a real number

To add or subtract complex numbers, do the following:

> Add or subtract the real and the imaginary parts separately.

Note: A real number is often called a scalar.

EXAMPLE

(i) If $u = 2 + 3i$ and $w = 1 - i$, express $2u + 3w$ in the form $a + bi$.

(ii) If $z_1 = 3 - 2i$ and $z_2 = 1 + i$, express $z_1 - 2z_2$ in the form $x + yi$.

Solution:

(i) $2u + 3w$
$= 2(2 + 3i) + 3(1 - i)$
$= 4 + 6i + 3 - 3i$
$= 4 + 3 + 6i - 3i$
$= 7 + 3i$

(ii) $z_1 - 2z_2$
$= (3 - 2i) - 2(1 + i)$
$= 3 - 2i - 2 - 2i$
$= 3 - 2 - 2i - 2i$
$= 1 - 4i$

Exercise 3.3
Write each of the following in questions 1–10 in the form $a + bi$, where $a, b \in \mathbb{R}$.

1. $2 + 3i + 4 + 2i$
2. $5 + 8i - 2 - 4i$
3. $11 + 3i - 5 + 4i$
4. $7 + 5i - 9 - 2i$
5. $-3 - 7i + 5 + 11i$
6. $5 + 2i + 4i - 8$
7. $2(4 + i) + 3(2 + i)$
8. $5(3 + 2i) + 2(1 + 3i)$
9. $2(2 - 3i) - 3(5 - i) + 11 + 6i$
10. $-2(-5 + 3i) + 5(1 + i) + i$

If $u = 2 + i$, $w = 3 - 2i$ and $z = 1 + i$, express questions 11–19 in the form $p + qi$, where $p, q \in \mathbb{R}$.

11. $u + w$
12. $u + z$
13. $w + z$
14. $u - i$
15. $w + 2i$
16. $w + 1$
17. $2u + z$
18. $w + 2z$
19. $3z - 2w - 3$

If $z_1 = 2 + 3i$, $z_2 = 1 - 5i$, $z_3 = 4$ and $z_4 = 2i$, express questions 20–25 in the form $x + yi$, where $x, y \in \mathbb{R}$.

20. $z_1 + z_2$
21. $z_2 + z_3$
22. $z_1 + z_4$
23. $2z_1 + z_2$
24. $2(z_3 + z_4)$
25. $z_3(z_2 - z_1)$

Multiplication of complex numbers

Multiplication of complex numbers is performed using the usual algebraic method, except:

i^2 is replaced with -1.

EXAMPLE

(i) Simplify $4(2 - i) + i(3 + 5i)$ and write your answer in the form $x + yi$, where $x, y \in \mathbb{R}$.

(ii) If $u = 1 - 3i$ and $w = 2 + i$, express uw in the form $p + qi$, where $p, q \in \mathbb{R}$.

Solution:

(i) $4(2 - i) + i(3 + 5i)$
$= 8 - 4i + 3i + 5i^2$
$= 8 - 4i + 3i + 5(-1)$
(replace i^2 with -1)
$= 8 - 4i + 3i - 5$
$= 3 - i$

(ii) uw
$= (1 - 3i)(2 + i)$
$= 1(2 + i) - 3i(2 + i)$
$= 2 + i - 6i - 3i^2$
$= 2 + i - 6i - 3(-1)$
(replace i^2 with -1)
$= 2 + i - 6i + 3$
$= 5 - 5i$

Exercise 3.4

Express questions 1–15 in the form $a + bi$, where $a, b \in \mathbb{R}$ and $i^2 = -1$.

1. $2i(3 - 2i)$
2. $3i(4 + 2i)$
3. $i(-1 + 2i)$
4. $i(2 - 5i) + i - 1$
5. $2(3 - i) + i(4 + 5i)$
6. $4(2 - i) + i(3 + 5i)$
7. $(2 + 3i)(4 + i)$
8. $(3 + i)(2 + i)$
9. $(3 + 2i)(2 - 5i)$
10. $(6 - i)(4 - 3i)$
11. $(-3 - 4i)(2 - i)$
12. $(2 + 3i)(2 - 3i)$
13. $(2 + i)^2$
14. $(3 - 2i)^2$
15. $(5 + 2i)^2$

If $u = 2 + 3i$ and $w = 1 - i$, express questions 16–19 in the form $p + qi$, where $p, q \in \mathbb{R}$ and $i^2 = -1$.

16. uw
17. u^2
18. $2iu^2$
19. $w^2 - 2w$

If $z_1 = 1 + i$, $z_2 = -1 + 2i$ and $z_3 = 2 + 3i$, express questions 20–23 in the form $x + yi$, where $x, y \in \mathbb{R}$ and $i^2 = -1$.

20. $z_1 z_2$
21. $z_2 z_3$
22. $z_1 z_3$
23. $i z_1^2$

24. If $u = 2 + 3i$, where $i^2 = -1$, show that $u^2 - 4u + 13 = 0$.

Complex conjugate

Two complex numbers that differ only in the sign of their imaginary parts are called **conjugate complex numbers**, each being the conjugate of each other. For example, $-3 + 4i$ and $-3 - 4i$ are complex conjugates and each is called the conjugate of the other.

If z represents a complex number, the conjugate of z is denoted by $\bar{z}$ (pronounced 'z bar').

$$z = a + bi \quad \Leftrightarrow \quad \bar{z} = a - bi$$

To find the conjugate, simply **change the sign of the imaginary part only**. Do **not** change the sign of the real part.

For example, if $z = -2 - 5i$, then $\bar{z} = -2 + 5i$.

Note: If a complex number is added to or multiplied by its conjugate, the result will **always** be a real number. If complex conjugates are subtracted, the result is **always** an imaginary number.

EXAMPLE

If $z = -2 + 3i$, simplify the following.

(i) $z + \bar{z}$ (ii) $z - \bar{z}$ (iii) $z\bar{z}$

Solution:

If $z = -2 + 3i$, then $\bar{z} = -2 - 3i$ (change sign of imaginary part only).

(i) $z + \bar{z}$
$= (-2 + 3i) + (-2 - 3i)$
$= -2 + 3i - 2 - 3i$
$= -4$ (real number)

(ii) $z - \bar{z}$
$= (-2 + 3i) - (-2 - 3i)$
$= -2 + 3i + 2 + 3i$
$= 6i$ (imaginary number)

(iii) $z\bar{z}$
$= (-2 + 3i)(-2 - 3i)$
$= -2(-2 - 3i) + 3i(-2 - 3i)$
$= 4 + 6i - 6i - 9i^2$
$= 4 - 9(-1)$
$= 4 + 9$
$= 13$ (real number)

Exercise 3.5

Find $\bar{z}$ for each of the following in questions 1–8.

1. $z = 3 + 2i$
2. $z = 4 - 3i$
3. $z = -2 + 6i$
4. $z = -3 - 7i$
5. $z = 1 - 5i$
6. $z = -1 + 3i$
7. $z = -4 - 5i$
8. $z = -2 + 3i$
9. Let $u = (4 - 3i) - (-3 - 6i)$. Express $\bar{u}$ in the form $a + bi$, where $a, b \in \mathbb{R}$.
10. Let $w = (1 + i)(2 - i)$. Express $\bar{w}$ in the form $p + qi$, where $p, q \in \mathbb{R}$.
11. Let $z = (1 + 3i)^2$. Express $\bar{z}$ in the form $x + yi$, where $x, y \in \mathbb{R}$.

Find (i) $z + \bar{z}$ (ii) $z - \bar{z}$ (iii) $z\bar{z}$ for each of the following in questions 12–15.

12. $z = 4 + 5i$
13. $z = 3 - 2i$
14. $z = -4 + 2i$
15. $z = -1 - i$
16. Let $z = 5 - 3i$. Find the real number k such that $k(z + \bar{z}) = 20$.
17. Let $u = 4 + i$. Find the real number p such that $p(z - \bar{z}) = 12i$.
18. $u = 2 + 3i$, where $i^2 = -1$. Evaluate $u + \bar{u} + u\bar{u}$.
19. If $w = 3 - 4i$, where $i^2 = -1$, show that $w^2 - 6w + w\bar{w} = 0$.
20. Let $u = 2 + i$ and $w = 3 - i$. Show that:
    (i) $u\bar{w} + \bar{u}w$ is a real number
    (ii) $u\bar{w} - \bar{u}w$ is an imaginary number
21. If $u = 3 + 2i$, evaluate $\sqrt{u^2 + \bar{u}^2}$.

COMPLEX NUMBERS

Division by a complex number

> Multiply the top and bottom by the conjugate of the bottom.

This will convert the complex number on the bottom into a real number. The division is then performed by dividing the real number on the bottom into each part on the top.

EXAMPLE

Express $\dfrac{1+7i}{4+3i}$ in the form $a + bi$, where $a, b \in \mathbb{R}$ and $i^2 = -1$.

Solution:

$$\dfrac{1+7i}{4+3i} = \dfrac{1+7i}{4+3i} \times \dfrac{4-3i}{4-3i}$$ (multiply top and bottom by the conjugate of the bottom)

Top by the top
$= (1+7i)(4-3i)$
$= 1(4-3i) + 7i(4-3i)$
$= 4 - 3i + 28i - 21i^2$
$= 4 - 3i + 28i - 21(-1)$
$= 4 - 3i + 28i + 21$
$= 25 + 25i$

Bottom by the bottom
$= (4+3i)(4-3i)$
$= 4(4-3i) + 3i(4-3i)$
$= 16 - 12i + 12i - 9i^2$
$= 16 - 9(-1)$
$= 16 + 9$
$= 25$

$\therefore \dfrac{1+7i}{4+3i} = \dfrac{25+25i}{25} = \dfrac{25}{25} + \dfrac{25}{25}i = 1 + i$

Exercise 3.6

Express questions 1–12 in the form $a + bi$, where $a, b \in \mathbb{R}$ and $i^2 = -1$.

1. $\dfrac{2+10i}{3+2i}$
2. $\dfrac{7+4i}{2-i}$
3. $\dfrac{3+4i}{2+i}$
4. $\dfrac{1+7i}{1-3i}$
5. $\dfrac{19-4i}{3-2i}$
6. $\dfrac{7-i}{1+i}$
7. $\dfrac{11-7i}{2+i}$
8. $\dfrac{7-17i}{5-i}$
9. $\dfrac{8-4i}{2-i}$
10. $\dfrac{3-2i}{2+3i}$
11. $\dfrac{2+i}{1-i}$
12. $\dfrac{4-2i}{2+i}$

13. $u = 9 + 7i$ and $w = 2 + i$. Express $\dfrac{u}{w}$ in the form $p + qi$, where $p, q \in \mathbb{R}$ and $i^2 = -1$.

14. $z_1 = 8 - 2i$ and $z_2 = 1 + i$. Express $\dfrac{z_1}{z_2}$ in the form $x + yi$, where $x, y \in \mathbb{R}$ and $i^2 = -1$.

15. $u = 3 + 2i$, $v = -1 + i$ and $w = u - v - 2$, where $i^2 = -1$.

    Express in the form $a + bi$: **(i)** w **(ii)** $\dfrac{2u + v}{w}$.

16. Let $u = 2 - i$.

    (i) Express $\left(u + \dfrac{5}{u}\right)$ in the form $a + bi$, where $a, b \in \mathbb{R}$ and $i^2 = -1$.

    (ii) Hence, solve for k, $k\left(u + \dfrac{5}{u}\right) = 24$.

17. Let $z = \dfrac{1 + i}{1 - i}$. Evaluate z^2.

18. Let $u = 2 + i$, where $i^2 = -1$.

    (i) Investigate if $\dfrac{5}{u} = \bar{u}$.

    (ii) Express $\dfrac{u + 10i}{u}$ in the form $a + bi$, where $a, b \in \mathbb{R}$, and evaluate $\sqrt{a^2 + b^2}$.

    (iii) Show that $iu + \dfrac{u}{i} = 0$.

19. Find the real part and the imaginary part of $\dfrac{-4 + 7i}{1 + 2i}$.

Equality of complex numbers

If two complex numbers are equal, then **their real parts are equal and their imaginary parts are also equal.**

For example, if $a + bi = c + di$, then $a = c$ and $b = d$.

This definition is useful when dealing with equations involving complex numbers.

Equations involving complex numbers are usually solved with the following steps.

1. Remove the brackets (if any).
2. Put an R under the real parts and an I under the imaginary parts to identify them.
3. Let the real parts equal the real parts and the imaginary parts equal the imaginary parts.
4. Solve these resultant equations (usually simultaneous equations).

Note: If one side of the equation does not contain a real part or an imaginary part, it should be replaced with 0 or $0i$, respectively.

COMPLEX NUMBERS

EXAMPLE 1

(i) Let $w = 1 + i$. Express $\dfrac{6}{w}$ in the form $x + yi$, where $x, y \in \mathbb{Z}$ and $i^2 = -1$.

(ii) a and b are real numbers such that $a\left(\dfrac{6}{w}\right) - b(w + 1) = 3(w + i)$.

Find the value of a and the value of b.

Solution:

(i) $\dfrac{6}{w} = \dfrac{6}{1 + i}$ (multiply top and bottom by the conjugate of the bottom)

$= \dfrac{6}{1+i} \times \dfrac{1-i}{1-i} = \dfrac{6 - 6i}{1 - i + i - i^2} = \dfrac{6 - 6i}{1 + 1} = \dfrac{6 - 6i}{2} = 3 - 3i$

(ii) $a\left(\dfrac{6}{w}\right) - b(w + 1) = 3(w + i)$

$a(3 - 3i) - b(1 + i + 1) = 3(1 + i + i)$ $\quad\left(\text{put in } \dfrac{6}{w} = 3 - 3i \text{ and } w = 1 + i\right)$

$a(3 - 3i) - b(2 + i) = 3(1 + 2i)$

$3a - 3ai - 2b - bi = 3 + 6i$ (remove the brackets)

$\text{R}\ \text{I}\text{R}\ \text{I}\text{R}\text{I}$ (identify real and imaginary parts)

Real parts = Real parts   **Imaginary parts = Imaginary parts**

$3a - 2b\ \ =\ \ 3$ ① $-3a - b = 6$ ②

Now solve the simultaneous equations ① and ②.

$\begin{aligned}3a - 2b &= 3 \quad ① \\ -3a - b &= 6 \quad ② \\ \hline -3b &= 9 \quad \text{(add)} \\ 3b &= -9 \\ b &= -3\end{aligned}$ $\qquad$ Put $b = -3$ into ① or ②.
$3a - 2b = 3$ ①
$3a - 2(-3) = 3$
$3a + 6 = 3$
$3a = -3$
$a = -1$

Thus, $a = -1$ and $b = -3$.

107

EXAMPLE 2

$z_1 = 4 - 2i$, $z_2 = -2 - 6i$. If $z_2 - pz_1 = qi$, where $p, q \in \mathbb{R}$, find p and q.

Solution:

$$z_2 - pz_1 = qi$$

The right-hand side has no real part, hence a 0, representing the real part, should be placed on the right-hand side.

Now the equation is:

$z_2 - pz_1 = 0 + qi$ (put 0 in for real part)

$(-2 - 6i) - p(4 - 2i) = 0 + qi$ (substitute for z_1 and z_2)

$-2 - 6i - 4p + 2pi = 0 + qi$ (remove the brackets)

R   I   R   I   R   I   (identify real and imaginary parts)

Real parts = Real parts   **Imaginary parts = Imaginary parts**

$-2 - 4p = 0$ ① $-6 + 2p = q$ ②

Solve between the equations 1 and 2.

$-2 - 4p = 0$ ①   Substitute $p = -\tfrac{1}{2}$ into equation ②.

$-4p = 2$ $-6 + 2p = q$

$4p = -2$ $-6 + 2(-\tfrac{1}{2}) = q$

$p = -\tfrac{2}{4} = -\tfrac{1}{2}$ $-6 - 1 = q$

     $-7 = q$

Thus, $p = -\tfrac{1}{2}$, $q = -7$.

Exercise 3.7

Solve questions 1–10 for x and y, where $i^2 = -1$ and $x, y \in \mathbb{R}$.

1. $2x + 3yi = 10 + 15i$
2. $5x - 4yi = 30 + 12i$
3. $4x - 5 + 3yi + 4i = 3 + 7i$
4. $x + y + xi + 2yi = 8 + 13i$
5. $(2x + y) + (3x - y)i = 7 + 3i$
6. $2(x + yi) + 2 - 5i = 4 + 3xi$
7. $(4x - 2) + i(x - 4) = 2(2 - y + yi)$
8. $x(3 + 4i) + 2y(2 + 3i) = 2i$
9. $x(2 + 3i) + y(4 + 5i) - 19i = 16$
10. $2x + 5yi = (1 - 3i)(4 + 2i)$
11. Find real k and l, such that $k(3 - 2i) + l(i - 2) = 5 - 4i$.
12. Find real p and q, such that $2p - q + i(7i + 3) = 2(2i - q) - i(p + 3q)$.
13. $z_1 = 4 - 3i$ and $z_2 = 5 + 5i$. Find real k and t such that $k(z_1 + z_2) = 18 + i(t + 2)$.

14. Let $z = 5 - 3i$. Find the real number k such that $ki + 4z = 20$.
15. Let $z = 1 - 2i$. Find the real numbers k and t such that $kz + t\bar{z} = 2z^2$.
16. Let $u = 2 - 3i$, where $i^2 = -1$. If $u + 3(p + 2qi) = 5 + 9i$, find p and q, where $p, q \in \mathbb{R}$.
17. (i) Let $w = -2 + i$. Express w^2 in the form $a + bi$, where $a, b \in \mathbb{R}$.
    (ii) Hence, solve for real k and t, $kw^2 = 2w + 1 + ti$.
18. (i) Let $z = 1 + i$. Show that $\dfrac{z}{\bar{z}} = i$.
    (ii) Hence, solve $k\left(\dfrac{z}{\bar{z}}\right) + t\bar{z} = -3 - 4i$ for real k and t.
19. (i) Let $z_1 = 5 + 12i$ and $z_2 = 2 - 3i$. Show that $\dfrac{z_1}{z_2} = -z_2$.
    (ii) Hence, solve for real p and q, $\dfrac{z_1}{z_2} = p(q + i) + 1$.
20. (i) Let $w = 1 - i$. Express $\dfrac{4}{w}$ in the form $x + yi$, where $x, y \in \mathbb{R}$ and $i^2 = -1$.
    (ii) a and b are real numbers such that $a\left(\dfrac{4}{w}\right) - b(w + 1) = 2(w + 4)$. Find the value of a and the value of b.
21. Let $z = 6 - 4i$. Find the real numbers s and t such that $\dfrac{s + ti}{4 + 3i} = z$.
22. Solve $(x + 2yi)(1 - i) = 7 + 5i$ for real x and real y.
23. (i) Express $\dfrac{3 - 2i}{1 - 4i}$ in the form $x + yi$.
    (ii) Hence or otherwise, find the values of the real numbers p and q such that $p + 2qi = \dfrac{17(3 - 2i)}{1 - 4i}$.
24. Let $z = a + bi$, where $a, b \in \mathbb{R}$. Find the value of a and the value of b for which $3z - 10i = (2 - 3i)z$.

Quadratic equations with complex roots

When a quadratic equation cannot be solved by factorisation, the following formula can be used.

> The equation $ax^2 + bx + c = 0$ has roots given by:
> $$x = \frac{-b \pm \sqrt{b^2 - 4ac}}{2a}$$

Note: The whole of the top of the right-hand side, including $-b$, is divided by $2a$.

It is often called the **quadratic** or **$-b$ formula**.

If $b^2 - 4ac < 0$, then the number under the square root sign will be negative, and so the solutions will be complex numbers.

EXAMPLE 1

Solve the equation (i) $z^2 + 4z + 5 = 0$ (ii) $z^2 - 6z + 13 = 0$.
Write your answers in the form $x + yi$ where $x, y \in \mathbb{R}$.

Solution:

(i) $z^2 + 4z + 5 = 0$
$az^2 + bz + c = 0$
$a = 1, b = 4, c = 5$
$$z = \frac{-b \pm \sqrt{b^2 - 4ac}}{2a}$$
$$z = \frac{-4 \pm \sqrt{(4)^2 - 4(1)(5)}}{2(1)}$$
$$z = \frac{-4 \pm \sqrt{16 - 20}}{2}$$
$$z = \frac{-4 \pm \sqrt{-4}}{2}$$
$$z = \frac{-4 \pm 2i}{2}$$
$$z = -2 \pm i$$

(ii) $z^2 - 6z + 13 = 0$
$az^2 + bz + c = 0$
$a = 1, b = -6, c = 13$
$$z = \frac{-b \pm \sqrt{b^2 - 4ac}}{2a}$$
$$z = \frac{6 \pm \sqrt{(-6)^2 - 4(1)(13)}}{2(1)}$$
$$z = \frac{6 \pm \sqrt{36 - 52}}{2}$$
$$z = \frac{6 \pm \sqrt{-16}}{2}$$
$$z = \frac{6 \pm 4i}{2}$$
$$z = 3 \pm 2i$$

COMPLEX NUMBERS

Note: Notice in both solutions the roots occur in conjugate pairs. If one root of a quadratic equation with real coefficients is a complex number, then the other root must also be complex and the conjugate of the first.

i.e. if $3 - 4i$ is a root, then $3 + 4i$ is also a root
if $-2 - 5i$ is a root, then $-2 + 5i$ is also a root
if $a + bi$ is a root, then $a - bi$ is also a root

EXAMPLE 2

Verify that $-2 + 5i$ is a root of the equation $z^2 + 4z + 29 = 0$ and find the other root.

Solution:

Method 1

If $-2 + 5i$ is a root, then when z is replaced by $-2 + 5i$ in the equation, the equation will be satisfied, i.e.

$$(-2 + 5i)^2 + 4(-2 + 5i) + 29 = 0$$

Check:
$(-2 + 5i)^2 + 4(-2 + 5i) + 29$
$= (-2 + 5i)(-2 + 5i) + 4(-2 + 5i) + 29$
$= 4 - 10i - 10i - 25 - 8 + 20i + 29$
$= 33 - 33 + 20i - 20i$
$= 0$

∴ $-2 + 5i$ is a root and
$-2 - 5i$ is the other root
(the conjugate of $-2 + 5i$)

Method 2

$z^2 + 4z + 29 = 0$

$a = 1, b = 4, c = 29$

$z = \dfrac{-b \pm \sqrt{b^2 - 4ac}}{2a}$

$z = \dfrac{-4 \pm \sqrt{(4)^2 - 4(1)(29)}}{2(1)}$

$z = \dfrac{-4 \pm \sqrt{16 - 116}}{2}$

$z = \dfrac{-4 \pm \sqrt{-100}}{2}$

$z = \dfrac{-4 \pm 10i}{2}$

$z = -2 \pm 5i$

∴ $-2 + 5i$ is a root and
$-2 - 5i$ is the other root

Sometimes we have to find unknown coefficients.

EXAMPLE 3

Let $z = 3 - 4i$ be one root of the equation $z^2 + pz + q = 0$, where $p, q \in \mathbb{R}$.
Find the value of p and the value of q.

Solution:
If $3 - 4i$ is a root, then $3 + 4i$ is also a root (roots occur in conjugate pairs).

Method: Form a quadratic equation with roots $3 - 4i$ and $3 + 4i$.

Let $z = 3 - 4i$ and $z = 3 + 4i$
$\therefore$ $z - 3 + 4i = 0$ and $z - 3 - 4i = 0$
And $(z - 3 + 4i)(z - 3 - 4i) = 0$ $\quad (0 \times 0 = 0)$
$z(z - 3 - 4i) - 3(z - 3 - 4i) + 4i(z - 3i - 4i) = 0$
$z^2 - 3z - 4zi - 3z + 9 + 12i + 4zi - 12i - 16i^2 = 0$
$z^2 - 6z + 9 - 16(-1) = 0$ $\quad (i^2 = -1)$
$z^2 - 6z + 25 = 0$

By comparing $z^2 - 6z + 25 = 0$ to $z^2 + pz + q = 0$
$p = -6$ and $q = 25$

Note: An alternative method is to substitute $3 - 4i$ into the equation $z^2 + pz + q = 0$, equate the coefficients and solve these equations.

Exercise 3.8

Solve each of the following equations in questions 1–12.

1. $x^2 - 6x + 13 = 0$
2. $z^2 - 2z + 10 = 0$
3. $x^2 + 4x + 5 = 0$
4. $z^2 + 10z + 34 = 0$
5. $z^2 + 4z + 13 = 0$
6. $x^2 - 10x + 41 = 0$
7. $z^2 + 2z + 2 = 0$
8. $x^2 - 2x + 5 = 0$
9. $x^2 + 8x + 17 = 0$
10. $x^2 + 4 = 0$
11. $x^2 + 25 = 0$
12. $z^2 + 9 = 0$

Solve for x in questions 13–15.

13. $2x^2 - 2x + 1 = 0$
14. $2x^2 - 6x + 5 = 0$
15. $4x^2 - 8x + 5 = 0$
16. Verify that $1 - 2i$ is a root of $z^2 - 2z + 5 = 0$ and find the other root.
17. Verify that $2 - 5i$ is a root of $z^2 - 4z + 29 = 0$ and find the other root.
18. Verify that $4 - 3i$ is a root of $x^2 - 8x + 25 = 0$ and find the other root.

19. Verify that $-1 - 2i$ is a root of $x^2 + 2x + 5 = 0$ and find the other root.
20. Verify that $-6 - i$ is a root of $z^2 + 12z + 37 = 0$ and find the other root.
21. p and k are real numbers such that $p(2 + i) + 8 - ki = 5k - 3 - i$.
    (i) Find the value of p and the value of k.
    (ii) Investigate if $p + ki$ is a root of the equation $z^2 - 4z + 13 = 0$.
22. (i) Express $\dfrac{1 + 7i}{1 - 3i}$ in the form $p + qi$.
    (ii) Hence, show that $\dfrac{1 + 7i}{1 - 3i}$ is a root of the equation $z^2 + 4z + 5 = 0$ and write down the other root in the form $a + bi$, where $a, b \in \mathbb{R}$.
23. Show that $\dfrac{11 - 7i}{2 + i}$ is a root of the equation $z^2 - 6z + 34 = 0$ and write down the other root in the form $p + qi$, where $p, q \in \mathbb{R}$.

Form a quadratic equation with roots for questions 24–29.

24. $-2 \pm i$
25. $1 \pm i$
26. $-1 \pm 3i$
27. $3 \pm 5i$
28. $\pm 4i$
29. $\pm i$
30. If $3 + 5i$ is a root of $x^2 + px + q = 0$, where $p, q \in \mathbb{R}$, find the value of p and q.
31. If $7 - i$ is a root of $z^2 + az + b = 0$, where $a, b \in \mathbb{R}$, find the value of a and b.
32. If $-3 - 3i$ is a root of $x^2 + mx + n = 0$, where $m, n \in \mathbb{R}$, find the value of m and n.
33. If $1 + 5i$ is a root of $x^2 + 2x + k = 0$, where $k \in \mathbb{R}$, find the value of k.

Proving that a line and a curve do not meet

In Chapter 1 on algebra we learned how to find the point(s) of intersection of a line and a curve. In this section we are going to show how to prove that a line and a curve do not meet.

To prove that a line and a curve do not intersect, do the following.

Write the quadratic equation in the form $ax^2 + bx + c = 0$.
Method 1
Using the formula $x = \dfrac{-b \pm \sqrt{b^2 - 4ac}}{2a}$, show that the quadratic equation $ax^2 + bx + c = 0$ has complex roots. Therefore, the line and curve do not intersect.

Method 2

Evaluate $b^2 - 4ac$.

If $b^2 - 4ac < 0$, then the quadratic equation $ax^2 + bx + c = 0$ has complex roots. Therefore, the line and the curve do not intersect.

EXAMPLE

Show that the line $x - y + 6 = 0$ does not meet the curve $x^2 + y^2 = 10$.

Solution:

$x - y + 6 = 0$ and $x^2 + y^2 = 10$

1. $x - y + 6 = 0$ (get x or y on its own from the line)
    $-y = -x - 6$
    $y = x + 6$ (y on its own)

2. $x^2 + y^2 = 10$

    $x^2 + (x + 6)^2 = 10$ (put in $(x + 6)$ for y)
    $x^2 + x^2 + 12x + 36 = 10$ ($(x+6)^2 = x^2 + 12x + 36$)
    $2x^2 + 12x + 36 = 10$ (simplify the left-hand side)
    $2x^2 + 12x + 26 = 0$ (subtract 10 from both sides)
    $x^2 + 6x + 13 = 0$ (divide by sides by 2 in the form $ax^2 + bx + c = 0$)

Method 1: Using the formula $x = \dfrac{-b \pm \sqrt{b^2 - 4ac}}{2a}$

$x = \dfrac{-6 \pm \sqrt{(6)^2 - 4(1)(13)}}{2(1)}$ ($a = 1, b = 6, c = 13$)

$x = \dfrac{-6 \pm \sqrt{36 - 52}}{2(1)} = \dfrac{-6 \pm \sqrt{-16}}{2} = \dfrac{-6 \pm 4i}{2} = -3 \pm 2i$

As the roots are complex, the line does not intersect the curve.

Method 2: Evaluate $b^2 - 4ac$

$b^2 - 4ac = (6)^2 - 4(1)(13) = 36 - 52 = -16 < 0$

As $b^2 - 4ac < 0$, the line does not intersect the curve.

Exercise 3.9

In questions 1–14, that the line *l* and the curve *c* do not intersect.

1. $l: y = 2x - 5,\quad c: y = x^2 - 2x + 8$
2. $l: y = -x - 5,\quad c: y = x^2 + 5x + 20$
3. $l: y = x - 3,\quad c: y = x^2 + 3x + 2$
4. $l: y = 3x - 11,\quad c: y = x^2 - 5x + 30$
5. $l: y = 1,\quad c: y = x^2 + 10$
6. $l: y = 3x - 2,\quad c: y = x^2 + 3x + 2$
7. $l: x - y + 4 = 0,\quad c: xy + 29 = 0$
8. $l: x + y - 6 = 0,\quad c: xy - 34 = 0$
9. $l: x + y - 2 = 0,\quad c: xy = 2$
10. $l: x - y - 1 = 0,\quad c: xy + 2 = x - 15$
11. $l: y = x + 4,\quad c: x^2 + y^2 = 6$
12. $l: y = x + 6,\quad c: x^2 + y^2 = 16$
13. $l: y = 8 - x,\quad c: x^2 + y^2 = 30$
14. $l: y = 10 - x,\quad c: x^2 + y^2 = 42$

15. (i) Show that $(x + 3)^2 = x^2 + 6x + 9$.
    (ii) l is the line $x - y + 3 = 0$ and c is the curve $y^2 = 2(x - 10)$. Verify that l and c do not intersect.

16. (i) Show that $(2y + 15)^2 = 4y^2 + 60y + 225$.
    (ii) l is the line $x - 2y - 15 = 0$ and c is the circle $x^2 + y^2 = 25$. Show l and c on a coordinate diagram.
    (iii) Verify algebraically that l and c do not intersect.

17. The numbers x and y are such that when they are added the result is 4 and when they are multiplied the result is 13. Show that x and y are not real numbers.

18. The numbers a and b are such that when they are added the result is 10 and when they are multiplied the result is 29. Show that a and b are not real numbers.

Argand diagram

An Argand diagram is used to plot complex numbers. It is very similar to the *x*- and *y*-axes used in coordinate geometry, except that the **horizontal** axis is called the **real axis (Re)** and the **vertical** axis is called the imaginary axis **(Im)**. It is also called the **complex plane**.

Each complex number must be written in the form $a + bi$ and then plot the point (a, b).

For example, the complex number $5 - 4i$ is represented by the point $(5, -4)$.

EXAMPLE

$z = 2 + 3i$. Plot z, $z - 3$, iz, $-i\bar{z}$ and $\dfrac{13}{z}$ on an Argand diagram.

Solution:
First write each complex number in the form $a + bi$.

$z = 2 + 3i = (2, 3)$

$z - 3 = 2 + 3i - 3 = -1 + 3i = (-1, 3)$

$iz = i(2 + 3i) = 2i + 3i^2 = 2i + 3(-1) = 2i - 3 = -3 + 2i = (-3, 2)$

$-i\bar{z} = -i(2 - 3i) = -2i + 3i^2 = -2i + 3(-1) = -2i - 3 = -3 - 2i = (-3, -2)$

$\dfrac{13}{z} = \dfrac{13}{2 + 3i} = \dfrac{13}{2 + 3i} \cdot \dfrac{2 - 3i}{2 - 3i} = \dfrac{26 - 39i}{4 - 6i + 6i - 9i^2} = \dfrac{26 - 39i}{13} = 2 - 3i = (2, -3)$

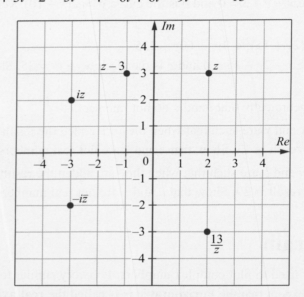

Exercise 3.10

1. Plot each of the following complex numbers on the Argand diagram below.

$z_1 = 5 + 2i$

$z_2 = -6 + 3i$

$z_3 = 3 - 4i$

$z_4 = -4 - 2i$

$z_5 = -3i$

$z_6 = 4$

$z_7 = 2(3 - i) + i(4 + 5i)$

$z_8 = 4(2 - i) + i(3 + 5i)$

$z_9 = 3(2 - 3i) + i(5 + 6i)$

$z_{10} = (3 - i)(-1 + i)$

$z_{11} = \dfrac{5i}{1 + 2i}$

$z_{12} = \dfrac{-1 - 9i}{1 + i}$

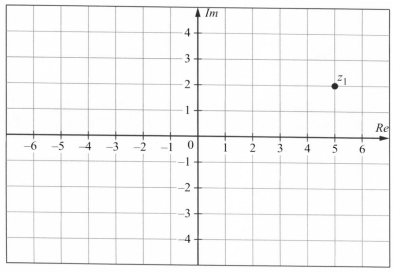

2. Two complex numbers are $u = 3 + 2i$ and $v = -1 + i$, where $i^2 = -1$.

   (i) Given that $w = u - v - 2$, show that $w = 2 + i$.

   (ii) Construct an Argand diagram with the range on the real axis from -5 to 7 and the range on the imaginary axis from -4 to 5. On this Argand diagram, plot each of the following complex numbers.

   (a) u (b) v (c) w (d) $2u$ (e) $4v$ (f) $-u$

   (g) iw (h) w^2 (i) v^2 (j) $\dfrac{u+v+w}{4}$ (k) $\dfrac{13}{w+2i}$ (l) $\dfrac{2u+v}{w}$

3. $u = 3 + 2i$, $v = 2 - 4i$ and $w = 4u + kv$. Find the real number k such that w lies on
   (i) the imaginary axis (ii) the real axis. In each case, plot w.

Modulus

The **modulus** of a complex number is the **distance** from the origin to the point representing the complex number on the Argand diagram.

If $z = a + bi$, then the modulus of z is written as $|z|$ or $|a + bi|$.

The point z represents the complex number $a + bi$.

The modulus of z is the distance from the origin, O, to the complex number $a + bi$.

Using Pythagoras' theorem, $|z| = \sqrt{a^2 + b^2}$.

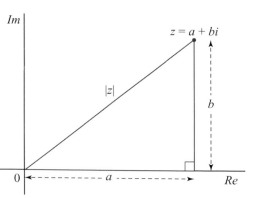

If $z = a + bi$, then
$|z| = |a + bi| = \sqrt{a^2 + b^2}$.

Notes:
1. i **never** appears when the modulus formula is used.
2. The modulus of a complex number is **always positive**.
3. Before using the formula, a complex number must be in the form $a + bi$.

EXAMPLE 1

Find **(i)** $|4 + 3i|$ **(ii)** $|5 - i|$ **(iii)** $|2i|$.

Solution:

(i) $|4 + 3i|$
$= \sqrt{4^2 + 3^2}$
$= \sqrt{16 + 9}$
$= \sqrt{25} = 5$

(ii) $|5 - i|$
$= \sqrt{5^2 + 1^2}$
$= \sqrt{25 + 1}$
$= \sqrt{26}$

(iii) $|2i| = |0 + 2i|$
$= \sqrt{0^2 + 2^2}$
$= \sqrt{0 + 4}$
$= \sqrt{4} = 2$

EXAMPLE 2

Let $u = 3 - 6i$ where $i^2 = -1$. Calculate $|u + 2i|$.

Solution:

$u = 3 - 6i$
$u + 2i = 3 - 6i + 2i$
$u + 2i = 3 - 4i$

$|u + 2i|$
$= |3 - 4i|$
$= \sqrt{3^2 + 4^2}$
$= \sqrt{9 + 16} = \sqrt{25} = 5$

EXAMPLE 3

Let $z_1 = 2 + 3i$ and $z_2 = 5 - i$.
Investigate whether $|z_1 + z_2| > |z_1 - z_2|$.

Solution:

$$z_1 = 2 + 3i \quad \text{and} \quad z_2 = 5 - i$$

$z_1 + z_2$
$= (2 + 3i) + (5 - i)$

$z_1 - z_2$
$= (2 + 3i) - (5 - i)$

$$= 2 + 3i + 5 - i$$
$$= 7 + 2i$$
$$\therefore |z_1 + z_2|$$
$$= |7 + 2i| = \sqrt{7^2 + 2^2} = \sqrt{49 + 4} = \sqrt{53}$$
$$= 7 \cdot 3 \text{ (correct to one decimal place)}$$

$$= 2 + 3i - 5 + i$$
$$= -3 + 4i$$
$$\therefore |z_1 - z_2|$$
$$= |-3 + 4i| = \sqrt{3^2 + 4^2} = \sqrt{9 + 16} = \sqrt{25} = 5$$

$$7 \cdot 3 > 5$$
$$\therefore |z_1 + z_2| > |z_1 - z_2|$$

EXAMPLE 4

For what values of k is $|11 + ki| = |10 - 5i|$, where $k \in \mathbb{Z}$?

Solution:
Given:
$$|11 + ki| = |10 - 5i|$$
$$\therefore \sqrt{11^2 + k^2} = \sqrt{10^2 + 5^2}$$
$$\sqrt{121 + k^2} = \sqrt{100 + 25}$$
$$\sqrt{121 + k^2} = \sqrt{125}$$
$$121 + k^2 = 125 \qquad \text{(square both sides)}$$
$$k^2 = 4$$
$$k = \pm\sqrt{4}$$
$$k = \pm 2$$

Exercise 3.11

Evaluate each of the following in questions 1–16.

1. $|3 + 4i|$
2. $|6 + 8i|$
3. $|5 - 12i|$
4. $|-8 - 15i|$
5. $|10 - 24i|$
6. $|-24 - 7i|$
7. $|-20 + 21i|$
8. $|-9 - 40i|$
9. $|3i|$
10. $|-2 + i|$
11. $|-5 - 6i|$
12. $|(3 - 2i)^2|$
13. $\left|\dfrac{-14 - 2i}{1 + i}\right|$
14. $\left|\dfrac{2 + i}{1 + 2i}\right|$
15. $\left|\dfrac{3 + i}{2 - i}\right|$
16. $\left|\dfrac{5 + i}{1 - i}\right|$

17. Let $u = 5 + 8i$ and $w = 3 + 7i$. Show that $|u + w| = 17$.
18. Evaluate $|4(2 - i) + i(3 + 5i)|$.
19. $u = 5 - 3i$ and $w = 3 - 6i$ are two complex numbers. Verify that $|u - 1| = |w + 2i|$.
20. Let $z_1 = 1 + 7i$ and $z_2 = 4 + 3i$.

    Express $\dfrac{z_1}{z_2}$ in the form $a + bi$, where $a, b \in \mathbb{R}$. Calculate $\left|\dfrac{z_1}{z_2}\right|$.
21. Let $u = 3 + 4i$ and $w = 12 - 5i$. Investigate whether $|u| + |w| = |u + w|$.
22. Let $z_1 = 2 + 3i$ and $z_2 = 5 - i$. Plot z_1, z_2 and $z_1 - z_2$ on an Argand diagram. Investigate whether $|z_1 + z_2| > |z_1 - z_2|$.
23. Let $w = 1 + 3i$. Investigate whether $|iw + w| = |iw| + |w|$.
24. (i) Let $u = (1 - 3i)(2 + i)$. Express u in the form $p + qi$, where $p, q \in \mathbb{Z}$.
    (ii) Verify that $|u + \bar{u}| = |u - \bar{u}|$.
25. Let $w = 2 + 5i$.
    (i) Express w^2 in the form $x + yi$, where $x, y \in \mathbb{R}$.
    (ii) Verify that $|w^2| = |w|^2$.
26. Let $z_1 = 1 + 7i$ and $z_2 = 1 - 3i$.

    (i) Express $z_1 z_2$ and $\dfrac{z_1}{z_2}$ in the form $a + bi$.

    (ii) Show that (a) $|z_1| \cdot |z_2| = |z_1 z_2|$ (b) $\dfrac{|z_1|}{|z_2|} = \left|\dfrac{z_1}{z_2}\right|$.

27. $u = 6 + 8i$ and $w = -5 + 12i$ are two complex numbers. Find the value of the real number k such that $k|u| = |w|$.
28. Let $z = 3 - 4i$. (i) Calculate $|z|$.
    (ii) Find the real numbers p and q such that $|z|(p + qi) + (q - pi) = 17 + 7i$.
    (iii) Find the real numbers s and t such that $|z|(s + ti) = \dfrac{5}{z}$.
29. $u = -5 + 12i$ and $w = 8 + 10i$ are two complex numbers. Which complex number is nearer to the origin? Justify your answer.
30. (i) Find the values of the real numbers x and y such that $3x + i(7 - 2y) = xi + 2(y + 3) - yi$.
    (ii) Explain why the complex numbers $x + yi$ and $y + xi$ are the same distance from the origin.
31. (i) Let $z = 1 + 7i$ and $w = -1 + i$. Express $\dfrac{z}{w}$ in the form $a + bi$, $a, b \in \mathbb{R}$ and $i^2 = -1$.

    (ii) Verify that $\dfrac{|z|}{|w|} = \left|\dfrac{z}{w}\right|$. (iii) Solve for real h and k: $hz = \left|\dfrac{z}{w}\right|kw + 16i$.
32. Let $u = 1 + 2i$. Solve for real a and b, $(1 + 2i)(a + bi) = |u|^2$.
33. Let $u = \dfrac{2 - i}{1 - 2i}$. Show that (i) $u = \dfrac{4}{5} + \dfrac{3}{5}i$ (ii) $|u| = 1$.
34. (i) If $x^2 = 9$, verify that $x = \pm 3$. (ii) If $|a + 3i| = 5$, $a \in \mathbb{R}$, find two possible values of a.
35. If $|8 + ki| = 10$, $k \in \mathbb{R}$, find two possible values of k.

36. If $|4 + qi| = |2 - 4i|$, $q \in \mathbb{R}$, find two possible values of q.
37. If $|a + ai| = |1 - 7i|$, $a \in \mathbb{R}$, find two possible values of a.
38. Let $z_1 = 8 + i$ and $z_2 = k + 7i$, $k \in \mathbb{R}$.
    If $|z_2| = |z_1|$, find two possible values of k.

Higher powers of i

Every integer (positive or negative whole number) power of i is a number of the set $\{1, -1, i, -i\}$.

$i = \sqrt{-1}$
$i^2 = -1$
$i^3 = i^2 \times i = (-1)i = -i$
$i^4 = i^2 \times i^2 = (-1)(-1) = 1$

$i = \sqrt{-1}$
$i^2 = -1$
$i^3 = -i$
$i^4 = 1$

$i^{4n} = 1$ (n is a multiple of 4)
$i^{4n+1} = i$ (n is one more than a multiple of 4)
$i^{4n+2} = -1$ (n is two more than a multiple of 4)
$i^{4n+3} = -i$ (n is three more than a multiple of 4)

EXAMPLE

Simplify the following.
(i) i^8 (ii) i^7 (iii) $-i(i^4 + i^5 + i^6)$ (iv) $(2i)^3$

Solution:
(i) $i^8 = i^4 \times i^4 = 1 \times 1 = 1$
(ii) $i^7 = i^4 \times i^3 = 1 \times -i = -i$
(iii) $\quad -i(i^4 + i^5 + i^6)$
$= -i(1 + i - 1)$
$= -i(i)$
$= -i^2 = -(-1) = 1$

$i^4 = 1$
$i^5 = i^4 \times i^1 = 1 \times i = i$
$i^6 = i^4 \times i^2 = 1 \times -1 = -1$

(iv) $(2i)^3 = 2i \times 2i \times 2i = 8i^3 = 8(-i) = -8i$

Exercise 3.12

Express questions 1–12 without indices.

1. i^2
2. i^3
3. i^4
4. i^5
5. i^6
6. i^7
7. i^8
8. i^{10}
9. i^{13}
10. i^{16}
11. i^{22}
12. i^{23}

Simplify questions 13–14.

13. $-2i(i^4 + i^5 + i^6)$
14. $3i(i^5 + i^6 + i^7)$

15. Express each of the following in the form $a + bi$.
    (i) $i^2 + 5i^3$
    (ii) $2i^6 + 3i^5$
    (iii) $2i^7 - 6i^{10}$

16. If $u = 2i$, verify that $u^3 + u^2 + 4u + 4 = 0$.

17. If $w = 3i$, verify that $w^3 - w^2 + 9w - 9 = 0$.

18. $z = \dfrac{1 + i}{1 - i}$. Calculate: (i) z^2 (ii) z^3 (iii) z^4 (iv) $-z(z^2 + z^3 + z^4)$

19. Let $u = \dfrac{3 + 4i}{4 - 3i}$. Evaluate: (i) u^2 (ii) u^3 (iii) u^4 (iv) $u(u^6 + u^7 + u^8)$

Further geometrical properties of complex numbers

Rotations

A **rotation** turns a point through an angle about a fixed point.

An **anticlockwise** turn is described as a **positive rotation**

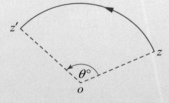

Written R_θ

A **clockwise** turn is described as a **negative rotation**

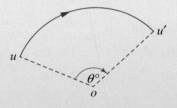

Written $R_{-\theta}$

Successive multiplication by i on an Argand diagram

Multiplication by i rotates a complex number by 90° ($R_{90°}$).
Multiplication by i^2 rotates a complex number by 180° ($R_{180°}$).
Multiplication by i^3 rotates a complex number by 270° ($R_{270°}$).
Multiplication by i^4 rotates a complex number by 360° ($R_{360°}$).

Note: Multiplication by $-i$, $-i^2$, $-i^3$ and $-i^4$ reverses the direction of the rotation.

If you multiply a complex number by a scalar (a number), then its modulus (distance from the origin) will be multiplied by this scalar. In other words, if you multiply a complex number by 3, then its distance from the origin will be three times as far from the origin as the original complex number. If you multiply a complex number by $\frac{1}{2}$, then its distance from the origin will be half as far from the origin as the original complex number.

> If z is a complex number, then $|kz| = k|z|$, where k is a real number.

If you multiply a complex number by $2i$, its modulus will be doubled and it will also be rotated by 90°.

EXAMPLE

(i) $z = 3 + 2i$. Represent z, iz, i^2z and i^3z on an Argand diagram.

(ii) $u = 2 + i$. Represent $2u$ on an Argand diagram. Verify that $|2u| = 2|u|$.

Solution:

(i) $i^2 = -1$ and $i^3 = -i$

$z = 3 + 2i$ (3, 2)

$iz = i(3 + 2i) = 3i + 2i^2 = 3i + 2(-1) = 3i - 2 = -2 + 3i$ (−2, 3)

$i^2z = -1(3 + 2i) = -3 - 2i$ (−3, −2)

$i^3z = -i(3 + 2i) = -3i - 2i^2 = -3i - 2(-1) = -3i + 2 = 2 - 3i$ (2, −3)

(ii) $u = 2 + i$

$2u = 2(2 + i) = 4 + 2i$ (4, 2)

(i)

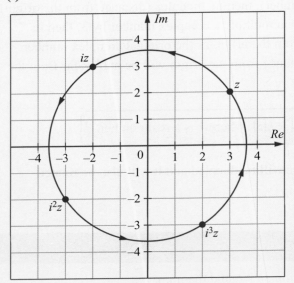

(ii)

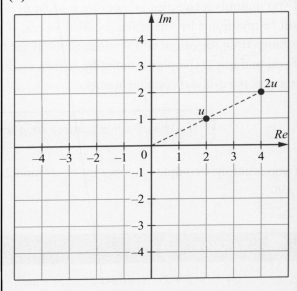

$|z| = |3 + 2i| = \sqrt{3^2 + 2^2} = \sqrt{9 + 4} = \sqrt{13}$
$|z| = |iz| = |i^2 z| = |i^3 z|$
In other words, they are all the same distance from the origin. This can be seen on the diagram as all four complex numbers lie on the same circle of centre O and radius $\sqrt{13}$.

$|u| = |2 + i| = \sqrt{2^2 + 1^2} = \sqrt{4 + 1} = \sqrt{5}$
$|2u| = |4 + 2i| = \sqrt{4^2 + 2^2} = \sqrt{16 + 4} = \sqrt{20}$
$\sqrt{20} = \sqrt{4 \times 5} = \sqrt{4}\sqrt{5} = 2\sqrt{5}$
$\therefore |2u| = 2|u|$
In other words, $2u$ is twice as far from the origin as u. This can be seen on the diagram. A line from the origin passes through u and $2u$.

Exercise 3.13

1. $z = 3 + 4i$. Investigate whether $|z| = |iz|$. Give a geometrical interpretation for your answer.

2. $z = 2 + i$ and $u = -4 + 3i$.

   (i) On the Argand diagram, plot:

   (a) z, iz, i^2z and i^3z

   (b) u, iu, i^2u and i^3z

   (ii) Describe the position of i^4z.

   (iii) Describe a transformation, which is not a rotation, that maps iz onto i^3z.

   (iv) Name the image of u under a central symmetry in iz.

   (v) Explain why iz and i^5z can be represented by the same point on the Argand diagram.

   (vi) $w = 2iu$. Give a geometrical interpretation of how to find the position of w on the Argand diagram.

   (vii) Using the diagram, explain why $|u| > |z|$.

   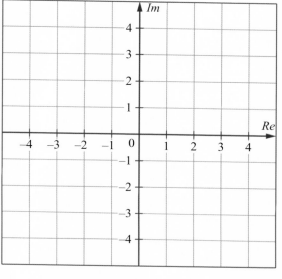

3. (i) Let $z = 7 + i$, $w = 3 - i$, $u = z - w$ and $v = \dfrac{z}{w}$.

   Express u and v in the form $a + bi$ and plot u and v on an Argand diagram.

   (ii) $u = kv$, where $k \in N$. Find the value of k.

   (iii) $v = lu$, where $l \in Q$. Find the value of l.

   (iv) On your Argand diagram, plot $2i^3v$ and $-iv$. Comment on the position of $2i^3v$ and $-iv$.

4. (i) $w = 3 - 2i$. On an Argand diagram, plot w, $-iw$, $-i^2w$, $-i^3w$ and $-i^4w$.

   (ii) Explain why w and i^4w have the same position on the Argand diagram.

   (iii) Is $i^2w = -i^2w$? Justify your answer.

   (iv) $u = 3iw$. Give a geometrical interpretation of u.

5. (i) Construct an Argand diagram with the real and imaginary axis from -4 to 4. Let $z_1 = 2 + 4i$. Verify that z_1 is a solution of the equation $z^2 - 4z + 20 = 0$ and write down z_2, the other solution.

   (ii) Plot z_1 and z_2 on the Argand diagram.

   (iii) Is $iz_1 = z_2$? Justify your answer.

   (iv) Describe the transformation that maps z_1 onto z_2.

   (v) Investigate whether the image of z_2 under an axial symmetry in the imaginary axis has the same position as i^2z_1 on the Argand diagram.

6. (i) Let $z = 2 + i$ and $w = -1 - 3i$.
   Plot z and w on the Argand diagram.
   (ii) Verify that $|w - z| = 5$.
   (iii) Draw the set k of all complex numbers such that each is a distance of 5 from z.
   (iv) What geometrical figure is represented by k?
   (v) Let $u = -2 + 4i$. Investigate whether $u \in k$.
   (vi) Write in the form $a + bi$, the image of w under a central symmetry in z.

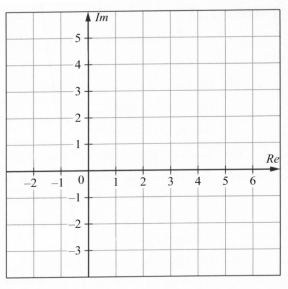

7. (i) Let $u = -2 + 3i$ and $v = 4 + i$. Find, in the form $a + bi$, the midpoint of $[uv]$.
   (ii) A circle is drawn with $a + bi$ as the centre and $\sqrt{10}$ as the radius. Show that the circle passes through u and v.

8. Four complex numbers $z_1, z_2, z_3,$ and z_4 are shown on the Argand diagram. They satisfy the following conditions:

   $z_2 = iz_1$
   $z_3 = kz_1$, where $k \in \mathbb{R}$
   $z_4 = z_2 + z_3$

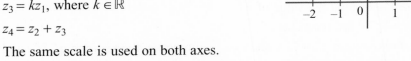

   The same scale is used on both axes.
   (i) Identify which number is which by labelling the points on the diagram.
   (ii) Write down the value of k.

9. (i) z is the complex number $1 + i$, where $i^2 = -1$. Find z^2 and z^3.
   (ii) Verify that $z^4 = -4$.   (iii) Show z, z^2, z^3 and z^4 on the Argand diagram.
   (iv) Make one observation about the pattern of points on the diagram.
   (v) Using the value of z^4, or otherwise, find the values of z^8, z^{12} and z^{16}.
   (vi) Based on the pattern of values in part (v) or otherwise, state whether z^{40} is positive or negative. Explain how you got your answer.
   (vii) Write z^{40} as a power of 2.   (viii) Find z^{41}.
   (ix) On an Argand diagram, how far from the origin is z^{41}?

CHAPTER 4: PERIMETER, AREA AND VOLUME

Perimeter and area

Formulae required (see the book of formulae and tables)

1. **Rectangle**

   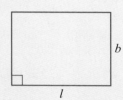

   Area = lb
   Perimeter = $2l + 2b = 2(l + b)$

2. **Square**

   Area = l^2
   Perimeter = $4l$

3. **Triangle**

   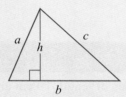

   Area = $\frac{1}{2}bh$
   Perimeter = $a + b + c$

4. **Triangle**

   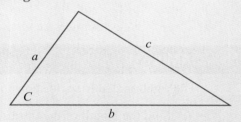

   Area = $\frac{1}{2}ab \sin C$
   Area = $\sqrt{s(s-a)(s-b)(s-c)}$
   Taking $s = \dfrac{a + b + c}{2}$

5. **Parallelogram**

   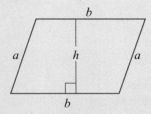

   Area = bh
   Perimeter = $2a + 2b$
   $\quad\quad\quad\;\, = 2(a + b)$

6. **Trapezium, a quadrilateral with two parallel sides of unequal length**

   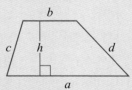

   Area = $\left(\dfrac{a + b}{2}\right)h$
   Perimeter = $a + c + b + d$

7. **Circle (disc)**

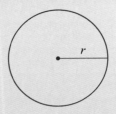

Area = πr^2
Circumference = $2\pi r$ = perimeter

8. **Sector of a circle**

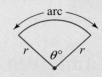

Area = $\dfrac{\theta}{360} \times \pi r^2$

Length of arc = $\dfrac{\theta}{360} \times 2\pi r$

$\left(\text{Similar to circle with } \dfrac{\theta}{360} \text{ in front of formulae}\right)$

Perimeter = $r + \dfrac{\theta}{360}(2\pi r) + r$

$= 2r + \dfrac{\theta}{360}(2\pi r)$

EXAMPLE 1

The right-angled triangle shown in the diagram has sides of length 10 cm and 24 cm.

(i) Find the length of the third side.
(ii) Find the length of the perimeter of the triangle.

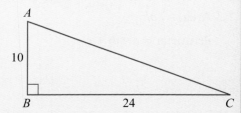

Solution:

(i) Using Pythagoras' theorem
$|AC|^2 = |AB|^2 + |BC|^2$
$|AC|^2 = 10^2 + 24^2$
$|AC|^2 = 100 + 576 = 676$
$|AC| = \sqrt{676} = 26$ cm

(ii) Perimeter = $|AB| + |BC| + |CA|$ = 10 + 24 + 26 = 60 cm

PERIMETER, AREA AND VOLUME

EXAMPLE 2

The figure is made up of a semicircle and a triangle (all dimensions are in cm).
Find the area of the figure in cm². (Assume $\pi = \frac{22}{7}$.)

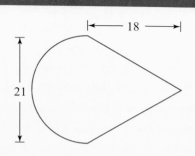

Solution:
Split the figure up into regular shapes, for which we have formulae to calculate the area.
Find the area of each shape separately and add these results together.

1. Area of semicircle
$= \frac{1}{2}\pi r^2$
$= \frac{1}{2} \times \frac{22}{7} \times \frac{21}{2} \times \frac{21}{2}$
$= 173 \cdot 25$ cm²

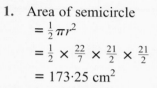 (radius $= \frac{21}{2}$)

2. Area of triangle
$= \frac{1}{2}bh$
$= \frac{1}{2} \times 21 \times 18$
$= 189$ cm²

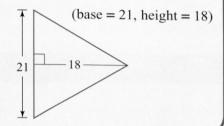 (base = 21, height = 18)

Area of figure = area of semicircle + area of triangle
$= 173 \cdot 25 + 189 = 362 \cdot 25$ cm²

EXAMPLE 3

The diagram represents a sector of a circle of radius 15 cm.
$|\angle POQ| = 72°$. Find:

(i) The area of the sector OPQ, in terms of π
(ii) The perimeter of the sector (assume $\pi = 3 \cdot 14$)

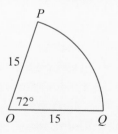

Solution:

(i) Area of sector

$$= \frac{\theta}{360} \times \pi r^2$$

$$= \frac{72}{360} \times \pi \times 15 \times 15$$

$$= \tfrac{1}{5} \times \pi \times 15 \times 15$$

$$= 45\pi \text{ cm}^2$$

(ii) Length of arc PQ

$$= \frac{\theta}{360} \times 2\pi r$$

$$= \frac{72}{360} \times 2 \times 3 \cdot 14 \times 15$$

$$= \tfrac{1}{5} \times 2 \times 3 \cdot 14 \times 15$$

$$= 18 \cdot 84 \text{ cm}$$

Perimeter $= 15 + 15 + 18 \cdot 84 = 48 \cdot 84$ cm

Notes: 1. When using $\pi = \frac{22}{7}$, it is good practice to write the radius as a fraction. For example, $21 = \frac{21}{1}$ or $4 \cdot 5 = \frac{9}{2}$.

2. If a question says 'give your answer in terms of π', then leave π in the answer: do not use $3 \cdot 14$ or $\frac{22}{7}$ or your calculator for π.

EXAMPLE 4

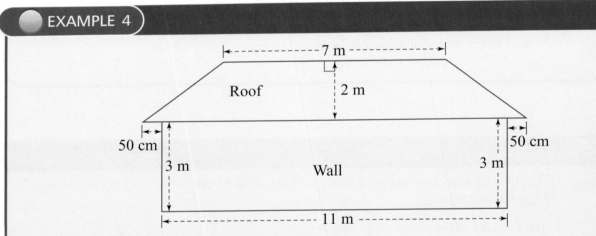

The diagram represents the frame of a bungalow.
Find the total area of the diagram in square metres.

Solution:

Note: 50 cm $= 0 \cdot 5$ m. It is vital that all calculations be done in the same units (in this case, metres).

Area of wall = lb
= (11)(3)
= 33 m²

Area of roof = $\left(\dfrac{a+b}{2}\right)h$

$a = 0\cdot 5 + 11 + 0\cdot 5 = 12$

$b = 7$

$h = 2$

Area of roof = $\left(\dfrac{12+7}{2}\right)2 = 19\,\text{m}^2$

Total area = 33 + 19 = 52 m²

EXAMPLE 5

Find the area of the triangle ABC:

(i) In the form $p\sqrt{p}$, where $p \in \mathbb{N}$
(ii) Correct to one decimal place

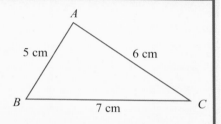

Solution:

(i) $s = \dfrac{5+6+7}{2} = 9$

Area △ABC = $\sqrt{s(s-a)(s-b)(s-c)}$
= $\sqrt{9(9-5)(9-6)(9-7)}$
= $\sqrt{(9)(4)(3)(2)}$
= $\sqrt{(36)(6)}$
= $6\sqrt{6}$ cm²

(ii) Area △ABC = $6\sqrt{6} = 6(2\cdot 0916) = 14\cdot 7$ cm²

Exercise 4.1

Unless otherwise stated, all dimensions are in cm and assume π = 3·14. All curved lines represent the circumference, or parts of the circumference, of a circle.

Find: (i) the perimeter (ii) the area of each of the following shapes in questions 1–9.

1.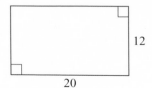

2.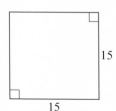

3.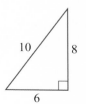

4. 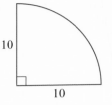 5. 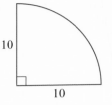 6.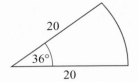

7. 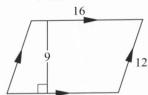 8. 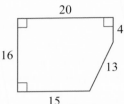 9.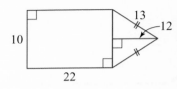

10. Calculate the area of the figure in the diagram.

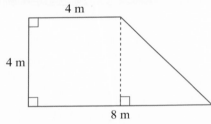

11. Calculate the area of the shaded region in the diagram.

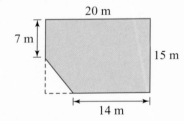

12. A right-angled triangle has sides of length 8 cm, 15 cm and 17 cm. Find its area.

13. The Department of the Environment designs a flag consisting of a blue triangle, OPQ, on a white background, $OCBA$, to display on beaches that have a very high standard of water purity.

    The flag $OABC$ is a rectangle with $|OA| = 1$ m and $|OC| = 80$ cm.

    P is the midpoint of $[CB]$ and Q is the midpoint of $[AB]$. Calculate the area of the blue section of the flag.

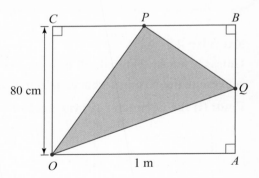

PERIMETER, AREA AND VOLUME

In questions 14–16, find the area of each of the following triangles (i) in surd form (ii) correct to one decimal place. All measurements are in cm.

14.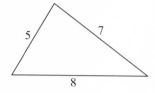

15.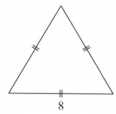

16.

Find the area of each of the following trapeziums in questions 17–19. All measurements are in cm.

17.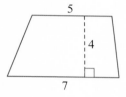

18.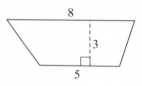

19.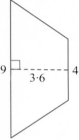

Calculate the area of the shaded region in questions 20–22, where *ABCD* is a rectangle and *PQRS* is a square.

20.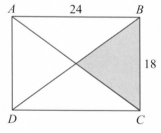

21.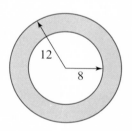

22.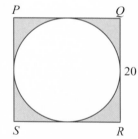

23. A rectangular piece of metal has a width of 16π cm. Two circular pieces, each of radius 8 cm, are cut from the rectangular piece, as shown.

    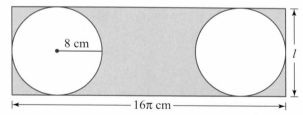

    (i) Find the length, *l*, of the rectangular piece of metal.

    (ii) Calculate the area of the metal not used (i.e. the shaded section), giving your answer in terms of π.

    (iii) Express the area of the metal not used as a percentage of the total area.

133

24. A rectangle has length 21 cm and width 20 cm.
    (i) Find the area of the rectangle.
    (ii) Find the length of the diagonal.

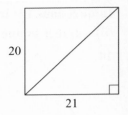

25. A circle is inscribed in a square, as shown. The radius of the circle is 9 cm.
    (i) Find the perimeter of the square.
    (ii) Calculate the area of the square.

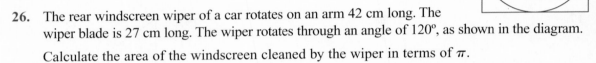

26. The rear windscreen wiper of a car rotates on an arm 42 cm long. The wiper blade is 27 cm long. The wiper rotates through an angle of 120°, as shown in the diagram. Calculate the area of the windscreen cleaned by the wiper in terms of π.

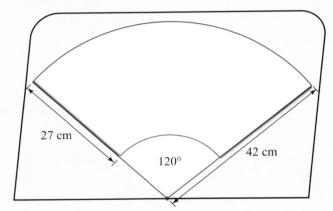

27. A circle with a radius of 7 cm has the trapezium *PQRS* inscribed as in diagram. Calculate the shaded area.

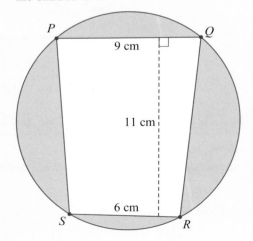

28. A circle with a radius of 3·5 cm is inscribed in the triangle *XYZ*, as shown in the diagram. Calculate the shaded area correct to the nearest integer, where $\pi = \frac{22}{7}$.

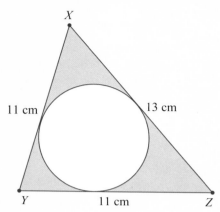

29. A running track is made up of two straight parts and two semicircular parts, as shown in the diagram. The length of each of the straight parts is 90 m. The diameter of each of the semicircular parts is 70 m. Calculate the length of the track correct to the nearest metre, taking $\pi = 3 \cdot 14$.

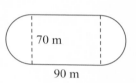

30. The diagram shows the perimeter of a running track consisting of two straight sections of length *l* and two semi-circular sections at each end of radius $\frac{100}{\pi}$ m, as shown.

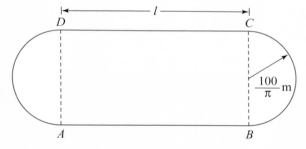

(i) Given that the perimeter of the track measures 400 m, find *l*.

(ii) A 1,500 m race starts at the point *A* and goes in the direction *ABCD*. At what point does the race finish?

(iii) An athlete completes this distance in 3 minutes 26 seconds. Find his average speed in m/s, correct to one decimal place.

31. The circle *p* has a radius of 4 cm. The circle *k* has a radius double the radius of the circle *p*. *k* and *p* touch at the point *W*. The circles are enclosed in the trapezium *ABCD*, as shown.
   $|AB| = 27$ cm and $|DC| = 18$ cm.
   Find the shaded area in the form $x - y\pi$, where $x, y \in N$.

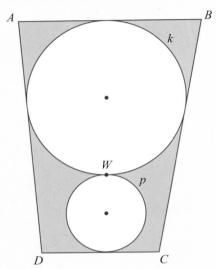

Given the perimeter or area

In some equations we are given the perimeter, the circumference or the area and asked to find missing lengths. Basically we are given **an equation in disguise** and we solve this equation to find the missing length.

EXAMPLE 1

The perimeter of a rectangle is 180 m. If length : breadth = 2 : 1, find the area of the rectangle.

Solution:
Let the length = $2x$ and the breadth = x.
Equation given in disguise:
$$\text{Perimeter} = 180$$
$$\therefore \quad 2x + x + 2x + x = 180$$
$$6x = 180$$
$$x = 30$$
$\therefore$ Breadth = 30 m
and length = $2x = 2(30) = 60$ m

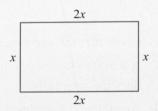

$$\text{Area} = l \times b$$
$$= 60 \times 30$$
$$= 1{,}800$$
$\therefore$ Area = $1{,}800$ m^2

PERIMETER, AREA AND VOLUME

EXAMPLE 2

The circumference of a circle is 37·68 cm. Calculate its area (assume $\pi = 3\cdot14$).

Solution:
Equation given in disguise:
Circumference = 37·68 cm
∴ $2\pi r = 37\cdot68$
$2(3\cdot14)r = 37\cdot68$
$6\cdot28r = 37\cdot68$
$r = \dfrac{37\cdot68}{6\cdot28} = 6$ cm

Area = πr^2
= $3\cdot14 \times 6 \times 6$
= 113·04
∴ Area of circle = 113·04 cm²

Exercise 4.2

If not given, drawing a rough diagram may help in solving the following problems.

1. The diagram shows a rectangle of length 42 cm.
   The area of the rectangle is 966 cm².
   (i) Find the height of the rectangle.
   (ii) Find the area of the shaded triangle.

   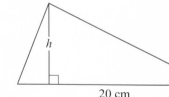
   42 cm

2. The area of a rectangle is 320 cm². If its length is 40 cm, calculate:
   (i) Its breadth   (ii) Its perimeter

3. (i) The perimeter of a square is 36 cm. Calculate its area.
   (ii) The area of a square is 25 cm². Calculate its perimeter.

4. The area of the triangle is 80 cm².
   The length of the base is 20 cm.
   Calculate its perpendicular height, h cm.

5. The triangle and the rectangle have equal area. Find h.

   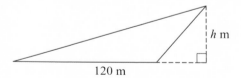
   h m
   120 m

   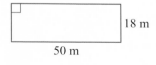
   18 m
   50 m

6. The length and breadth of a rectangle are in the ratio 3 : 2, respectively. The length of the rectangle is 12 cm. Find its breadth and its area.

7. The perimeter of a rectangle is 120 m. If length : breadth = 3 : 1, find the area of the rectangle.

137

8. The area of a rectangle is 128 m². If length : breadth = 2 : 1, find the length and the breadth of the rectangle.

The table below shows certain information on circles, including the value of π to be used. In each case, write down the equation given in disguise and use this to find the radius and complete the table.

	π	Circumference	Area	Radius
9.	π	10π		
10.	π		9π m²	
11.	π		6.25π cm²	
12.	$\frac{22}{7}$	264 cm		
13.	$\frac{22}{7}$		616 m²	
14.	3·14	157 mm		
15.	3·14		1256 m²	
16.	π		30.25π cm²	
17.	$\frac{22}{7}$		346·5 m²	
18.	3·14		452·16 cm²	

19. A piece of wire is 308 cm in length.

    308 cm

    The wire is bent into the shape of a circle.
    Calculate the radius of the circle.
    (Assume $\pi = \frac{22}{7}$.)

20. A piece of wire of length 66 cm is in the shape of a semicircle, as shown.
    Find the radius length of the semicircle.
    (Assume $\pi = \frac{22}{7}$.)

    66 cm

21. The semicircular shape shown in the diagram has a diameter of 16 cm. Taking $\pi = 3.14$:

    (i) Find the length of the perimeter of the shape, correct to the nearest centimetre
    (ii) Find the area of the shape, correct to the nearest square centimetre

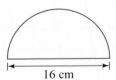

16 cm

PERIMETER, AREA AND VOLUME

22. The diagram shows a small circle drawn inside a larger circle. The small circle has an area of 25π cm². The larger circle has a circumference of 16π cm. Calculate the area of the shaded region in terms of π.

23. The area of the sector shown is 31·4 cm². Calculate the value of r. (Assume $\pi = 3\cdot14$.)

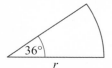

24. The area of the sector shown is 12·56 cm². Calculate the value of r. (Assume $\pi = 3\cdot14$.)

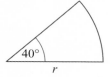

25. The area of the trapezium shown is 68·25 cm². Calculate the value of h.

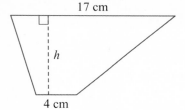

26. The area of the trapezium shown is 198 cm². Calculate the value of x.

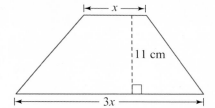

Volume and surface area

The **volume** of a solid is the amount of space it occupies.
Volume is measured in cubic units, such as cubic metres (m³) or cubic centimetres (cm³).
Capacity is the volume of a liquid or gas and is usually measured in litres.

Note: 1 litre = 1,000 cm³ = 1,000 ml

The **surface area** of a solid is the **total area of its outer surface**. It is measured in square units such as square metres or square centimetres.
To calculate the surface area of a solid you have to find the area of each face and add them together (often called the total surface area). With some objects, such as a sphere, the surface area is called the curved surface area.

Note: It is usual to denote volume by V and surface area by SA.

Rectangular solids

Formulae required

1. **Rectangular solid (cuboid)**

$$V = lbh$$

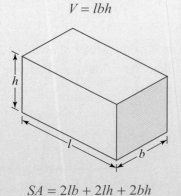

$$SA = 2lb + 2lh + 2bh$$

2. **Cube**

$$V = l^3$$

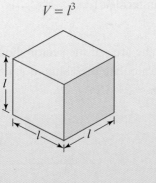

$$SA = 6l^2$$

EXAMPLE 1

The volume of a rectangular block is 560 cm³.
If its length is 14 cm and its breadth is 8 cm, find:

(i) Its height   (ii) Its surface area.

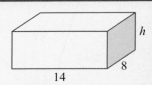

Solution:

(i) Equation given in disguise:
Volume = 560 cm³
$(14)(8)h = 560$
$112h = 560$
$h = \frac{560}{112} = 5$ cm

(ii) Surface area
$= 2lb + 2lh + 2bh$
$= 2(14)(8) + 2(14)(5) + 2(8)(5)$
$= 224 + 140 + 80$
$= 444$ cm²

EXAMPLE 2

The surface area of a cube is 96 cm².
Calculate its volume.

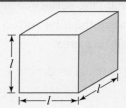

Solution:

Let the length of one side of the cube be l cm. Equation given in disguise:

Surface area = 96 cm²

$6l^2 = 96$

$l^2 = 16$

$l = 4$ cm

Volume = l^3

$= 4^3$

$= 64$ cm³

Thus, the volume of the cube is 64 cm³.

Exercise 4.3

In questions 1–3, find (i) the volume (ii) the surface area of a solid rectangular block with the following dimensions.

1. 6 cm, 5 cm, 4 cm
2. 12 m, 8 m, 6 m
3. 20 mm, 9 mm, 7 mm

4. The volume of a rectangular block is 480 cm³. Its length is 12 cm and its breadth is 8 cm. Calculate:

   (i) Its height (ii) Its surface area

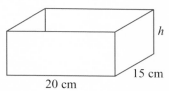

5. How many litres of water can be stored in a rectangular tank measuring 1·5 m by 70 cm by 50 cm?

   (**Note:** 1 litre = 1,000 cm³)

6. An open rectangular tank (no top) is full of water. The volume of water in the tank is 2·4 litres. If its length is 20 cm and its breadth is 15 cm, find:

   (i) Its height (ii) Its surface area
   (**Note:** 1 litre = 1,000 cm³)

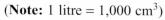

7. How many rectangular packets of tea measuring 12 cm by 4 cm by 4 cm can be packed into a cardboard box measuring 96 cm by 36 cm by 32 cm?

8. The volume of a cube is 27 cm³. Calculate its surface area.

9. The volume of a cube is 64 cm³. Calculate its surface area.

10. The surface area of a cube is 24 cm². Calculate its volume.

11. The surface area of a cube is 150 cm². Calculate its volume.

12. The sides of a rectangular block are in the ratio 2 : 3 : 7. If its volume is 2,688 cm³, find its dimensions and hence its surface area.

13. The surface area of a solid rectangular block is 258 cm². If its breadth is 6 cm and height is 5 cm, calculate: (i) Its length (ii) Its volume

14. A jeweller buys a rectangular block of gold of length 4 cm, width 3 cm and height 2 cm. 1 cm³ of gold costs €500.

    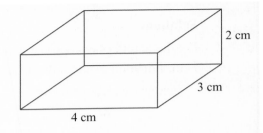

    (i) Calculate the cost of the block of gold.
    (ii) The jeweller needs 250 mm³ of gold to make a gold ring. How many rings can be made from the block?
    (iii) Each ring is sold for €150. Calculate the amount of profit the jeweller makes on each ring.

15. A solid rectangular metal block has length 12 cm and width 5 cm. The volume of the block is 90 cm³.

    (i) Find the height of the block in cm.
    (ii) Find the total surface area of the block in cm².
    (iii) Each cm³ of the metal has a mass of 8·4 g. The total mass of a number of these metal blocks is 113·4 kg. How many blocks are there?

Uniform cross-section

Many solid objects have the same cross-section throughout their length.

Here are some examples.

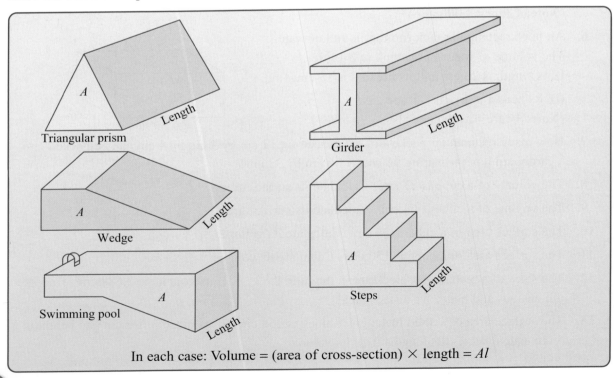

In each case: Volume = (area of cross-section) × length = Al

PERIMETER, AREA AND VOLUME

The above objects are called prisms. A prism is a solid object which has the same cross-section throughout its length and its sides are parallelograms.
A solid cylinder has a uniform cross-section, but it is not a prism.
To find the volume of a solid object with a uniform cross-section, find the area of the cross-section and multiply this by its length.

From the book of formulae and tables

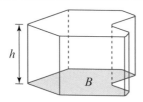

$V = Bh$

Solid of uniform cross-section (prism) taking B as the area of the base

EXAMPLE

The diagram shows the design of a swimming pool.
Calculate the capacity of the pool in m³.

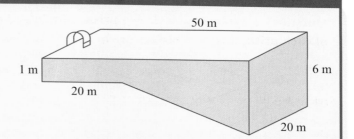

Solution:
The uniform cross-section is a combination of a rectangle and a triangle.

Area of cross-section
$= l \times b + \frac{1}{2}bh$
$= 50 \times 1 + \frac{1}{2} \times 5 \times 30$
$= 50 + 75 = 125 \text{ m}^2$

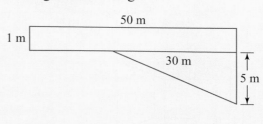

Volume = (area of cross-section) × width
$= 125 \times 20 = 2{,}500 \text{ m}^3$

∴ The capacity of the pool is 2,500 m³.

Exercise 4.4
Calculate the volume of each of the following solids in questions 1–6 (all dimensions are in cm).

1.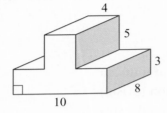

2.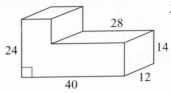

3.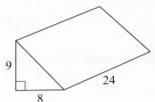

4.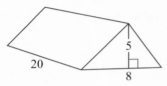

5.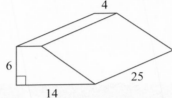

6.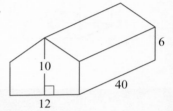

Questions 7, 8 and 9 each show a prism with one of the bases shaded. Calculate the volume of each prism. (All dimensions are in cm.)

7.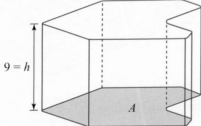

Area of base, $A = 14\cdot8$

8.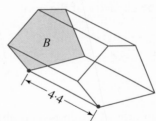

Area of base, $B = 27\cdot5$

9.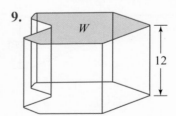

Area of base, $W = 93\cdot25$

10. The diagram shows a steel girder.
    (i) Calculate the area of its cross-section (shaded region).
    (ii) Calculate the volume, in cm³, of steel used to manufacture it.

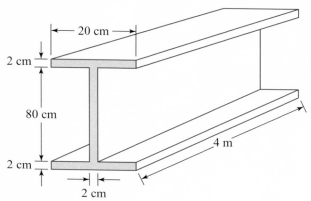

11. Five rectangular-shaped concrete steps are constructed as shown. Each step measures 1·2 m by 0·4 m and the total height is 1 m, with each step having the same height of 0·2 m. Calculate the volume of the solid concrete construction.

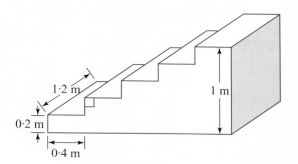

12. The diagram shows the design of a swimming pool.
    (i) Calculate the capacity of the pool in m³.
    (ii) Calculate the time, in hours and minutes, taken to fill the pool with water if the water is delivered by a pipe at the rate of 10 m³/min.
    (iii) Calculate the cost of heating the water for 15 hours if the average cost per cubic metre per hour is 0·08 c.

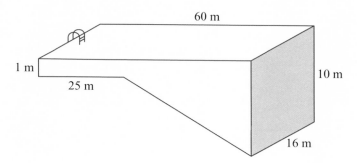

13. The diagram shows a triangular prism which has sloping sides that are perpendicular to each other.
    (i) Calculate the area of its cross-section (the shaded region).
    (ii) If its volume is 120 cm³, find x.

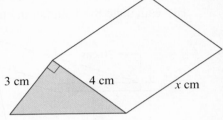

Cylinder, sphere and hemisphere

Formulae required (see the book of formulae and tables)

Cylinder:
Volume: $V = \pi r^2 h$
Curved surface area: $CSA = 2\pi rh$
Total surface area: $TSA = 2\pi rh + 2\pi r^2$

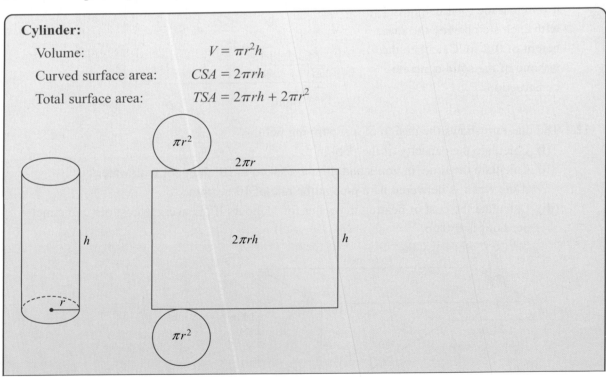

PERIMETER, AREA AND VOLUME

Sphere:

Volume: $\quad V = \frac{4}{3}\pi r^3$

Curved surface area: $\quad CSA = 4\pi r^2$

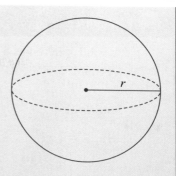

Hemisphere:

Volume: $\quad V = \frac{2}{3}\pi r^3$

Curved surface area: $\quad CSA = 2\pi r^2$

Total surface area: $\quad TSA = 2\pi r^2 + \pi r^2 = 3\pi r^2$

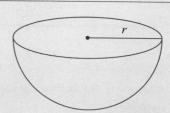

EXAMPLE 1

Find: **(i)** the volume **(ii)** the total surface area of a closed cylindrical can of radius 14 cm and height 10 cm (assume $\pi = \frac{22}{7}$).

Solution:

(i) $V = \pi r^2 h$

$V = \frac{22}{7} \times \frac{14}{1} \times \frac{14}{1} \times \frac{10}{1}$

$V = 6{,}160 \text{ cm}^3$

(ii) $TSA = 2\pi rh + 2\pi r^2$

$= \frac{2}{1} \times \frac{22}{7} \times \frac{14}{1} \times \frac{10}{1} + \frac{2}{1} \times \frac{22}{7} \times \frac{14}{1} \times \frac{14}{1}$

$= 880 + 1{,}232$

$= 2{,}112 \text{ cm}^2$

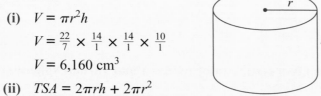

EXAMPLE 2

A solid sphere has a radius of 6 cm. Calculate: **(i)** its volume **(ii)** its curved surface area. (Assume $\pi = 3\cdot 14$.)

Solution:

(i) $V = \frac{4}{3}\pi r^3$
$= \frac{4}{3} \times 3\cdot 14 \times 6 \times 6 \times 6$
$= 904\cdot 32 \text{ cm}^3$

(ii) $CSA = 4\pi r^2$
$= 4 \times 3\cdot 14 \times 6 \times 6$
$= 452\cdot 16 \text{ cm}^2$

Exercise 4.5

Complete the following table, which gives certain information about various closed cylinders.

	π	Radius	Height	Volume	Curved surface area	Total surface area
1.	$\frac{22}{7}$	7 cm	12 cm			
2.	3·14	15 cm	40 cm			
3.	π	8 mm	11 mm			
4.	$\frac{22}{7}$	3·5 m	10 m			
5.	3·14	12 cm	40 cm			
6.	π	13 mm	30 mm			

Complete the following table, which gives certain information about various spheres.

	π	Radius	Volume	Curved surface area
7.	$\frac{22}{7}$	21 cm		
8.	3·14	9 m		
9.	π	6 mm		
10.	$\frac{22}{7}$	10·5 cm		
11.	3·14	7·5 cm		
12.	π	1·5 m		

PERIMETER, AREA AND VOLUME

Complete the following table, which gives certain information about various hemispheres.

	π	Radius	Volume	Curved surface area	Total surface area
13.	π	15 mm			
14.	π	$1\frac{1}{2}$ cm			
15.	$\frac{22}{7}$	42 cm			
16.	3·14	12 m			

17. A hollow plastic pipe has an external diameter of 16 cm and an internal diameter of 10 cm. Calculate the volume of plastic in 2 m of pipe. (Assume $\pi = 3\cdot14$.)

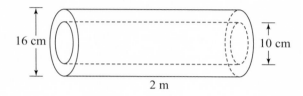

18. (i) The perimeter of a square lawn is 96 m. Find the area of the lawn in m².
    (ii) A garden roller in the shape of a cylinder has a diameter of 75 cm and is 1 m wide, as shown in the diagram. Calculate the curved surface area of the roller in m², correct to one decimal place.
    (iii) What percentage of the lawn will be rolled when the roller has completed 9 revolutions?

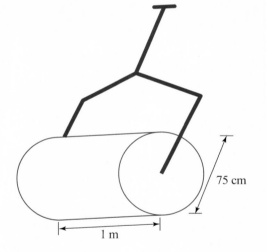

19. A cylindrical jug has a radius of 6 cm and a height of 40 cm. If the jug is full of lemonade, how many cylindrical tumblers, each with a radius of 4 cm and a height of 10 cm, can be filled from the jug?

20. A machine part consists of a hollow sphere floating in a closed cylinder full of oil. The height of the cylinder is 28 cm, the radius of the cylinder is 15 cm and the radius of the sphere is $\frac{21}{2}$ cm. Taking π to be $\frac{22}{7}$, find the volume of:
    (i) The cylinder
    (ii) The sphere
    (iii) The oil

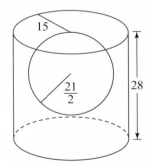

Cone

Formulae required (see the book of formulae and tables)

Volume:	$V = \frac{1}{3}\pi r^2 h$
Curved surface area:	$CSA = \pi r l$
Total surface area:	$TSA = \pi r l + \pi r^2$
Pythagoras' theorem:	$l^2 = r^2 + h^2$

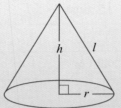

Notes: l is called the slant height.

A cone is often called a **right circular cone**, as its vertex is directly above the centre of the base and its height is at right angles to the base.

EXAMPLE 1

A right circular cone has a height of 12 cm and a base radius of 5 cm. Find:
(i) Its volume (ii) Its curved surface area (assume $\pi = 3\cdot 14$)

Solution:

(i) Volume of cone
$= \frac{1}{3}\pi r^2 h$
$= \frac{1}{3} \times 3\cdot 14 \times 5 \times 5 \times 12$
$= 314 \text{ cm}^3$

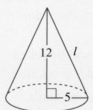

(ii) Slant height is missing
$l^2 = r^2 + h^2$
$l^2 = 5^2 + 12^2$
$l^2 = 25 + 144$
$l^2 = 169$
$l = 13$ cm

Curved surface area
$= \pi r l$
$= 3\cdot 14 \times 5 \times 13$
$= 204\cdot 1 \text{ cm}^2$

PERIMETER, AREA AND VOLUME

EXAMPLE 2

(i) A right circular cone has a height of 10 cm and a base radius of 18 cm. Find its volume in terms of π.

(ii) The cone is cut horizontally to its base at a height of 5 cm from its base into two sections, a cone A and a frustum B, as shown.
   (a) Find the radius of the cone A.
   (b) Find the volume of the cone A in terms of π.
   (c) Hence or otherwise, calculate the volume of the frustum, B, in terms of π.

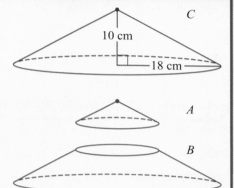

Solution:

(i) Volume of cone $C = \frac{1}{3}\pi r^2 h$
$= \frac{1}{3}\pi(18)^2(10)$
$= 1{,}080\pi$ cm^3

(ii) (a) We must use similar triangles to find the radius of cone A.
$\dfrac{r}{R} = \dfrac{h}{H}$
$\dfrac{r}{18} = \dfrac{5}{10}$
$10\,r = 90$
$r = 9$ cm

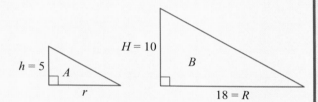

(b) Volume of cone $A = \frac{1}{3}\pi r^2 h$
$= \frac{1}{3}\pi(9)^2(5)$
$= 135\,\pi$ cm^3

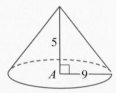

Method 1: Use subtraction

(c) Volume of frustum B = volume of cone C − volume of A

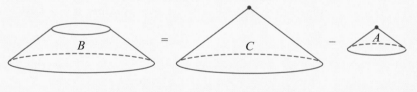

$= 1{,}080\,\pi \qquad -135\pi \qquad = 945\pi$ cm^3

Method 2: Use the formula for the volume of a frustum (see the book of formulae and tables)

$$V = \tfrac{1}{3}\pi h(R^2 + Rr + r^2)$$
$$= \tfrac{1}{3}\pi(5)[18^2 + 18(9) + (9)^2]$$
$$= \tfrac{1}{3}\pi(5)(324 + 162 + 81)$$
$$= \tfrac{1}{3}\pi(5)(567) = 945\pi \text{ cm}^3$$

Exercise 4.6
Complete the following table, which gives certain information about various cones.

	π	Radius	Height	Slant height	Volume	Curved surface area
1.	π	8 cm	6 cm			
2.	$\tfrac{22}{7}$		20 mm	29 mm		
3.	3·14	3 cm		5 cm		
4.	π	1·5 m		2·5 m		
5.	3·14		9 cm	41 cm		
6.	π	8 m		17 m		
7.	$\tfrac{22}{7}$	2·8 cm	4·5 cm			
8.	π	4·8 mm		5 mm		
9.	π	12 m	35 m			
10.	3·14	11 cm		61 cm		

11. A cone has a radius length of 7 cm and a height of 2·4 cm.

    Calculate: (i) Its volume (ii) Its total surface area (assume $\pi = \tfrac{22}{7}$)

12. A cone has a radius length of 18 cm and a height of 16·25 cm.

    Calculate: (i) Its volume (ii) Its total surface area (assume $\pi = 3\cdot 14$)

13. (i) A right circular cone *C*, with a height of 15 cm has a base with a radius of 12 cm. Find its volume in terms of π.
    (ii) The cone *C* is cut horizontally and divided into two sections, as in the diagram. Section *A* is a cone and section *B* is a frustum.
       (a) Write down the height of the cone *A*. Justify your answer for *h*.
       (b) Find, in terms of π, the volume of cone *A*.
       (c) Hence or otherwise, calculate the volume of the (frustum) section *B* in terms of π.

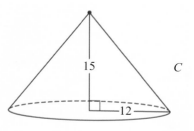
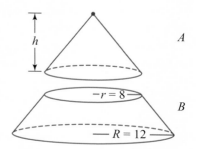

14. A right circular cone *S* has dimensions given in cm, as in the diagram. *S* is cut horizontally to its base and divided into two sections, *P* and *Q*.
    (i) Write down the radius *R* of the cone *S*. Justify your answer for *R*.
    (ii) Find the volume of *S*, correct to one decimal place, with $\pi = 3\cdot14$.
    (iii) Hence or otherwise, find the volume of the (frustum) section *Q*, correct to one decimal place, with $\pi = 3\cdot14$.

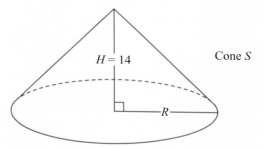
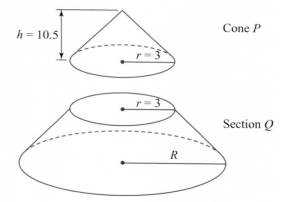

15. A right circular cone M has dimensions given in cm, as in the diagram. M is divided horizontally to its base into two sections, J and K.

    (i) Write down the radius, r, of the cone J. Justify your answer for r.

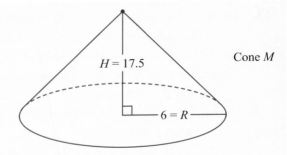
Cone M

(ii) Find the volume of J, taking $\pi = \frac{22}{7}$.

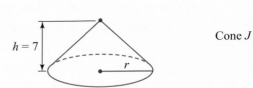
Cone J

(iii) Find the volume of the (frustum) section K, with $\pi = \frac{22}{7}$.

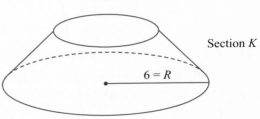
Section K

16. (i) A cone has radius x and height 3x. Find its volume in term of π and x.
    (ii) A second cone has twice the radius and half the height of the first cone. Find the ratio of the volume of the second cone to the volume of the first.

17. An egg timer consists of two identical cones of height 6 cm and base radius 4 cm. Sand occupies half the volume of one cone and flows from one to the other at a rate of $\frac{4\pi}{45}$ cm³ per second.
    (i) Calculate the volume of each cone in terms of π.
    (ii) Calculate the length of time it takes for the sand to flow from one cone into the other.

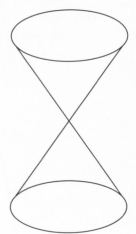

Compound volumes

Many of the objects we need to find the volume of will be made up of different shapes.
When this happens, do the following.

1. Split the solid up into regular shapes for which we have formulae to calculate the volume or surface area.
2. Add these results together (sometimes we subtract these results).

EXAMPLE 1

A solid object consists of three parts: a cone height 8 cm, A, a cylinder, B, and a hemisphere, C, each having a radius of 6 cm, as shown. If the height of the object is 26 cm, calculate its volume in terms of π.

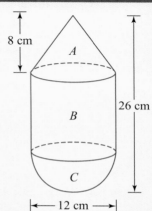

Solution:

A: Volume of cone $= \frac{1}{3}\pi r^2 h$
$= \frac{1}{3}\pi \times 6 \times 6 \times 8$
$= 96\pi$ cm^3

B: Volume of cylinder $= \pi r^2 h$
$= \pi \times 6 \times 6 \times 12$
$= 432\pi$ cm^3

C: Volume of hemisphere $= \frac{2}{3}\pi r^3$
$= \frac{2}{3}\pi \times 6 \times 6 \times 6$
$= 144\pi$ cm^3

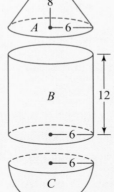

Total volume = volume of cone + volume of cylinder + volume of hemisphere
$= (96\pi + 432\pi + 144\pi)$ cm^3 $= 672\pi$ cm^3

EXAMPLE 2

An ashtray consists of a solid block with a cone excavated to contain the ash. The radius of the cone is 10 cm with a depth of 6 cm. The sides of the block touch the base circle of the cone, as in the diagram.

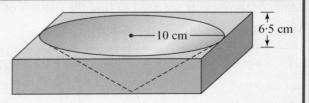

(i) Find the volume of the cone (take $\pi = 3 \cdot 14$).
(ii) Find the volume of the block.
(iii) Hence find the volume of material that makes up the ashtray.
(iv) Comment on the effectiveness or otherwise of this ashtray design. Suggest an alternative design from the shapes covered in this chapter.

Solution:

(i) Volume of cone $= \frac{1}{3}\pi r^2 h$
$= \frac{1}{3}(3 \cdot 14)(10)^2(6) = 628 \text{ cm}^3$

(ii) Volume of block $= lbh$
$= (20)(20)\left(6\frac{1}{2}\right) = 2{,}600 \text{ cm}^3$

(iii) Volume of ashtray material $= 2{,}600 - 628 = 1{,}972 \text{ cm}^3$

(iv) The useful space in this ashtray is 628 cm^3.
$\Rightarrow \frac{628}{2{,}600} \times 100 \approx 24\%$ of the total volume

It follows that 76% (100% − 24%) of this ashtray consists of material. In my opinion, this is too much material from both a practical and cost viewpoint. A good ashtray design should have more space than material.

In addition, the extremely pointed vertex at the base is not very effective and might be a safety issue (ash not properly crushed to extinguish the spark). My suggestion for a more efficient design would be a cylindrical space inside a solid cylinder, as in the diagram. You can make your own design.

PERIMETER, AREA AND VOLUME

Exercise 4.7

1. A glass container is in the shape of a cone surmounted by a cylinder, as shown. The height of the cylindrical part is 20 cm and the length of its radius is 8 cm. The slant height of the cone is 17 cm. Show that the volume of the container is $1,600\pi$ cm^3.

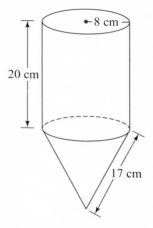

2. A test tube consists of a hemisphere of diameter 3 cm surmounted by a cylinder, as shown. The total height of the test tube is $16\frac{1}{2}$ cm. Calculate, in terms of π, the volume of the test tube.

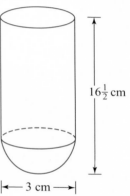

3. A boiler is in the shape of a cylinder with hemispherical ends, as shown in the diagram. The total length of the boiler is 30 m and its diameter is 12 m. Find in terms of π:
   (i) Its volume
   (ii) Its surface area

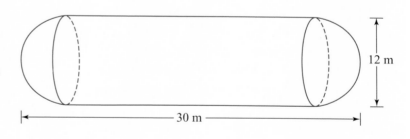

4. A solid object consists of three parts: a hemisphere, a cylinder and a cone, as shown, each having a diameter of 18 cm. If the height of the cone is 12 cm and the total height is 35 cm, calculate its volume and surface area in terms of π.

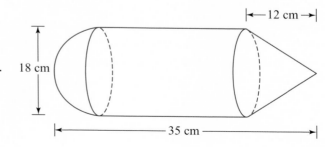

5. A buoy consists of an inverted cone surmounted by a hemisphere, as shown. If the radius of the hemisphere is 6 cm and the height of the cone is 9·1 cm, calculate, assuming $\pi = 3\cdot14$:
   (i) The volume of the buoy
   (ii) The surface area of the buoy

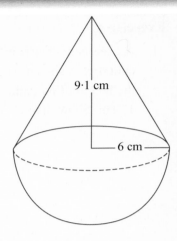

6. A tent has dimensions as shown in the diagram.

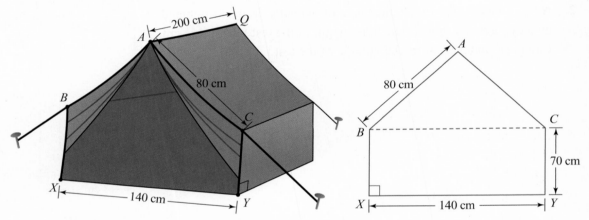

   (i) Find the area of the rectangular section $BXYC$.
   (ii) Show that the area of the triangular section $ABC = 700\sqrt{15}$ cm.
   (iii) Given that the tent has uniform length $= |AQ| = 200$ cm, calculate its volume to the nearest cm^3.

7. A wax crayon is in the shape of a cylinder of diameter 10 mm, surmounted by a cone of slant height 13 mm.
   (i) Show that the vertical height of the cone is 12 mm.
   (ii) Show that the volume of the cone is 100π mm^3.
   (iii) Given that the volume of the cylinder is 15 times the volume of the cone, find the volume of the crayon, in cm^3, correct to two decimal places.
   (iv) How many complete crayons like this one can be made from 1 kg of wax, given that each cm^3 of wax weighs 0·75 grams?

8. (i) A hot water container is in the shape of a hemisphere on top of a cylinder, as shown. The hemisphere has a radius of 25 cm and the container has a height of 100 cm. Find the internal volume of the container in litres, giving your answer correct to the nearest litre.

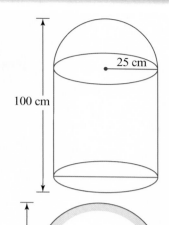

(ii) An external solid plastic lagging jacket 5 cm thick is constructed to fit exactly a round the hot water container and in the same shape. The jacket does not cover the bottom of the container. Find the total volume of the container, including the lagging jacket.

(iii) Hence, find the volume of plastic in the lagging jacket correct to the nearest litre.

(iv) Comment on the suggestion that the lagging jacket should cover the bottom of the container.

9. The dimensions of the gable end of a garage (the rectangle and triangle put together) are shown in the diagram.

   (i) Find the exterior area of the gable end.
   (ii) The length of the garage is 8·25 m. Find its volume.

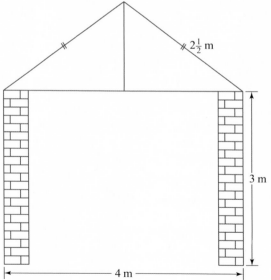

10. A solid sphere with a radius of 21 cm has a vertical cylindrical shaft with a radius of 1 cm drilled through its centre, c.

   (i) Calculate the volume remaining when the shaft is drilled. Take $\pi = \frac{22}{7}$.
   (ii) Show that Volume cylinder : Volume sphere = 1 : 294.

Note: The shaft is not an absolute cylinder – it has a slight bulge at each end. However, at the scale involved here it is not significant.

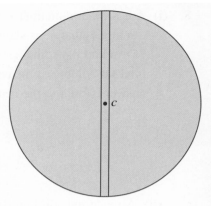

11.

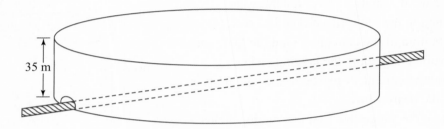

A stadium is designed in the shape of a cylinder. The stadium has a height of 35 cm and a radius of 100 m. A light rail line is constructed in a tunnel which passes directly through the centre, c, of the stadium on a ground-level track. The face of the tunnel is a semi-circle. The tunnel is half cylindrical in shape with a radius of 5 m.

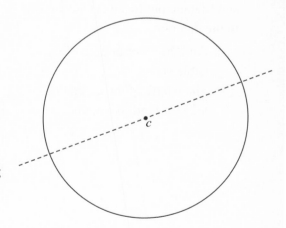

  (i) Find the volume of the stadium including the rail line in terms of π.
  (ii) Find the volume of the rail tunnel in terms of π.
  (iii) Hence, express the volume of the rail tunnel as a percentage of the volume of the stadium in the form $\frac{a}{b}\%$ where $a, b \in \mathbb{N}$.

Given the volume or surface area

In some questions we are given the volume or surface area and asked to find a missing dimension. As before, write down the **equation given in disguise** and solve this equation to find the missing dimension.

EXAMPLE

(i) A cylinder has a volume of 192π cm^3. If its radius is 4 cm, calculate its height.
(ii) The volume of a sphere is $\frac{32}{3}\pi$ cm^3. Calculate its radius.

Solution:

(i) Equation given in disguise:

$$\text{Volume of cylinder} = 192\pi \text{ cm}^3$$
$$\pi r^2 h = 192\pi$$
$$r^2 h = 192 \quad \text{(divide both sides by } \pi\text{)}$$
$$16h = 192 \quad \text{(put in } r = 4\text{)}$$
$$h = 12 \text{ cm} \quad \text{(divide both sides by 16)}$$

(ii) Equation given in disguise:

$$\text{Volume of sphere} = \frac{32}{3}\pi \text{ cm}^3$$
$$\tfrac{4}{3}\pi r^3 = \tfrac{32}{3}\pi$$
$$4\pi r^3 = 32\pi \quad \text{(multiply both sides by 3)}$$
$$4r^3 = 32 \quad \text{(divide both sides by } \pi\text{)}$$
$$r^3 = 8 \quad \text{(divide both sides by 4)}$$
$$r = 2 \text{ cm} \quad \text{(take the cube root of both sides)}$$

Exercise 4.8

1. A cylinder has a volume of 720π cm^3. If its radius is 6 cm, calculate:
    (i) Its height
    (ii) Its curved surface area in terms of π
2. The curved surface area of a sphere is 144π cm^2. Calculate:
    (i) Its radius
    (ii) Its volume in terms of π

3. The volume of a cone is 320π cm³. If the radius of the base is 8 cm, calculate its height.

4. The volume of a solid sphere is 36π cm³. Calculate:
   (i) Its radius
   (ii) Its surface area in terms of π

5. A solid cylinder has a volume of 96π cm³. If its height is 6 cm, calculate:
   (i) Its radius
   (ii) Its total surface area in terms of π

6. The curved surface area of a cylinder is 628 cm² and its radius is 5 cm. Calculate:
   (i) Its height
   (ii) Its volume (assume $\pi = 3.14$)

7. A solid cylinder has a volume of 462 m³. If the height is 12 m, assuming $\pi = \frac{22}{7}$, calculate:
   (i) Its radius
   (ii) Its total surface area

8. The curved surface area of a cone is 60π cm². If the radius of its base is 6 cm, calculate:
   (i) Its slant height
   (ii) Its volume in terms of π

9. A cone has a volume of $\frac{160}{3}\pi$ cm³. If the radius of the base is 4 cm, find its height.

10. The radius of a cylinder is 2·8 cm and its volume is 49·28 cm³. Calculate, assuming $\pi = \frac{22}{7}$:
    (i) Its height
    (ii) Its curved surface area

11. The volume of a solid cylinder is 401·92 m³. If its height is 8 m, calculate, assuming $\pi = 3.14$:
    (i) Its radius
    (ii) Its total surface area

12. The volume of a cone is $1,215\pi$ cm³. If the height is five times the radius of the base, calculate the height of the cone.

13. A buoy at sea is in the shape of a hemisphere with a cone on top, as in the diagram. The radius of the base of the cone is 0·9 m and its vertical height is 1·2 m.
    (i) Find the vertical height of the buoy.
    (ii) Find the volume of the buoy in terms of π.
    (iii) When the buoy floats, 0·8 m of its height is above water. Find, in terms of π, the volume of that part of the buoy that is above the water.

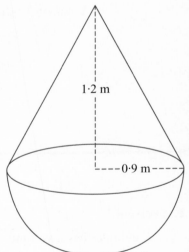

14. A fishing float consists of a cone on a hemisphere, each of radius 5 cm. The cone's volume is $1\frac{1}{5}$ times that of the hemisphere.
    (i) Find the volume of the hemisphere in terms of π.
    (ii) Find the volume of the cone in terms of π and h.
    (iii) Hence or otherwise, find the value of h.

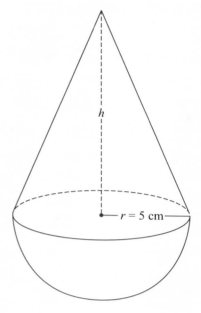

15. A team trophy for the winners of a football match is in the shape of a sphere supported on a cylindrical base, as shown. The diameter of the sphere and of the cylinder is 21 cm.
    (i) Find the volume of the sphere in terms of π.
    (ii) The volume of the trophy is $6{,}174\pi$ cm^3. Find the height of the cylinder.

16. (i) Write down, in terms of π, the volume of a hemisphere with a radius of length $4\frac{1}{2}$ cm.
    (ii) A pencil case in the shape of a cylinder with a hemisphere at each end is shown. The length of the case is 23 cm and the diameter is 9 cm. Calculate the volume of the pencil case in terms of π.

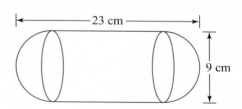

Equal volumes

Many questions involve equal volumes with a missing dimension. As before, write down the **equation given in disguise** and solve this equation to find the missing dimension.

Notes:
1. Moving liquid
In many questions we have to deal with moving liquid from one container to another container of different dimensions or shape. To help us solve the problem, we use the following fact:

> The volume of the moved liquid does not change.

2. Recasting
Many of the questions we meet require us to solve a recasting problem. What happens is that a certain solid object is melted down and its shape is changed. We use the following fact:

> The volume remains the same after it is melted down.

3. Displaced liquid
In many questions we have to deal with situations where liquid is displaced by immersing or removing a solid object. In all cases the following principle helps us to solve these problems:

> Volume of displaced liquid = volume of immersed or removed solid object

In problems on moving liquid or recasting or displaced liquid, it is good practice not to put in a value for π (i.e. do **not** put in $\pi = \frac{22}{7}$ or $\pi = 3\cdot14$), as the πs normally cancel when you write down the equation given in disguise.

 EXAMPLE 1

A sphere of radius 15 cm is made of lead. The sphere is melted down. Some of the lead is used to form a solid cone of radius 10 cm and height 27 cm. The rest of the lead is used to form a cylinder of height 25 cm. Calculate the length of the radius of the cylinder.

Solution:

Equation given in disguise:

Volume of cylinder + volume of cone = volume of sphere

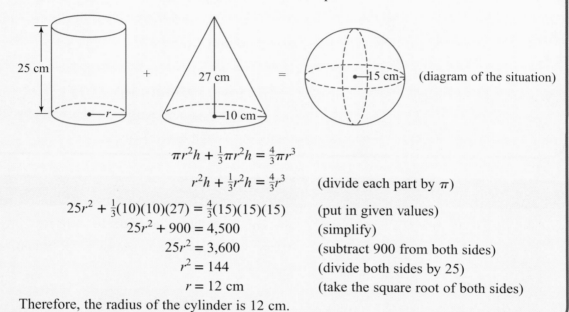 (diagram of the situation)

$$\pi r^2 h + \tfrac{1}{3}\pi r^2 h = \tfrac{4}{3}\pi r^3$$

$$r^2 h + \tfrac{1}{3} r^2 h = \tfrac{4}{3} r^3 \quad \text{(divide each part by } \pi)$$

$$25 r^2 + \tfrac{1}{3}(10)(10)(27) = \tfrac{4}{3}(15)(15)(15) \quad \text{(put in given values)}$$

$$25 r^2 + 900 = 4{,}500 \quad \text{(simplify)}$$

$$25 r^2 = 3{,}600 \quad \text{(subtract 900 from both sides)}$$

$$r^2 = 144 \quad \text{(divide both sides by 25)}$$

$$r = 12 \text{ cm} \quad \text{(take the square root of both sides)}$$

Therefore, the radius of the cylinder is 12 cm.

EXAMPLE 2

(i) Find, in terms of π, the volume of a solid metal sphere of radius 6 cm.

(ii) Five such identical spheres are completely submerged in a cylinder containing water. If the radius of the cylinder is 8 cm, by how much will the level of the water drop if the spheres are removed from the cylinder?

Solution:

(i) Volume of sphere $= \tfrac{4}{3}\pi r^3 = \tfrac{4}{3}\pi(6)(6)(6) = 288\pi \text{ cm}^3$

(ii) Diagram:

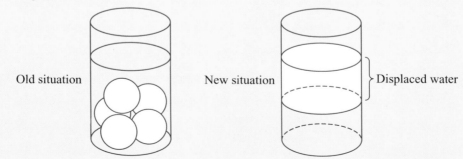

Equation given in disguise:

Volume of displaced water = volume of five spheres

Diagram:

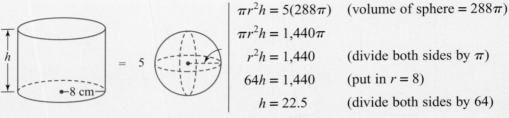

$\pi r^2 h = 5(288\pi)$ (volume of sphere = 288π)
$\pi r^2 h = 1{,}440\pi$
$r^2 h = 1{,}440$ (divide both sides by π)
$64h = 1{,}440$ (put in $r = 8$)
$h = 22.5$ (divide both sides by 64)

Thus, the level of water in the cylinder would fall by 22·5 cm.

EXAMPLE 3

(i) The volume of a hemisphere is 486π cm³. Find the radius of the hemisphere.

(ii) Find the volume of the smallest rectangular box that the hemisphere will fit into.

Solution:

(i) Volume $\tfrac{1}{2}$ sphere $= \tfrac{1}{2}\left[\tfrac{4}{3}\pi r^3\right]$

$486\pi = \tfrac{2}{3}\pi r^3$

$486 = \tfrac{2}{3} r^3$

$1{,}458 = 2r^3$

$729 = r^3$

$9 \text{ cm} = r$

(ii)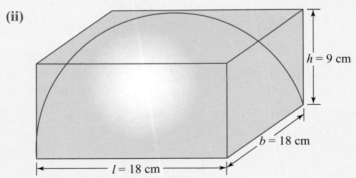

Volume rectangular box $= l \times b \times h$
$= 18 \times 18 \times 9$
$= 2{,}916 \text{ cm}^3$

Exercise 4.9

Find the missing dimensions in questions 1–6. In each case, the volumes are equal (all dimensions in centimetres).

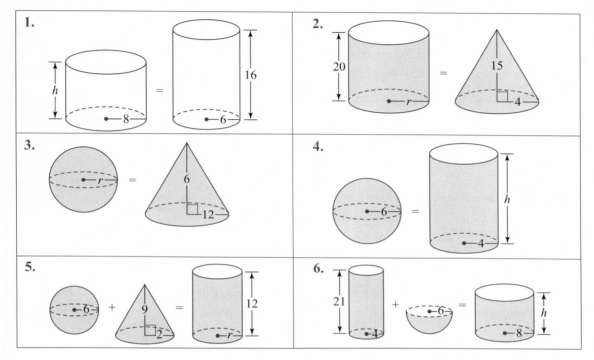

7. A solid lead cylinder of base radius 2 cm and height 15 cm is melted down and recast as a solid cone of base radius 3 cm. Calculate the height of the cone.

8. (i) A solid cylinder made of lead has a radius of length 15 cm and a height of 135 cm. Find its volume in terms of π.
   (ii) The solid cylinder is melted down and recast to make four identical right circular solid cones. The height of each cone is equal to twice the length of its base radius. Calculate the base radius length of the cones.

9. A cylinder of internal diameter 8 cm and height 18 cm is full of liquid. The liquid is poured into a second cylinder of internal diameter 12 cm. Calculate the depth of the liquid in this second cylinder.

10. A spherical golf ball has a diameter of 4 cm.
    (i) Find the volume of the golf ball in terms of π.
    (ii) A cylindrical hole on a golf course is 10 cm in diameter and 12 cm deep. The hole is half full of water. Calculate the volume of water in the hole in terms of π.
    (iii) The golf ball is dropped into the hole. Find the rise in the level of the water, correct to two decimal places.

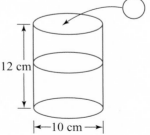

11. A solid metal rectangular block 30 cm by 24 cm by 15 cm is melted down and recast into cubes of side 3 cm. How many such cubes are made?
12. A solid is in the shape of a hemisphere surmounted by a cone, as in the diagram.
    (i) The volume of the hemisphere is 18π cm³.
        Find the radius of the hemisphere.
    (ii) The slant height of the cone is $3\sqrt{5}$ cm.
        Show that the vertical height of the cone is 6 cm.
    (iii) Show that the volume of the cone equals the volume of the hemisphere.
    (iv) This solid is melted down and recast in the shape of a solid cylinder. The height of the cylinder is 9 cm. Calculate its radius.

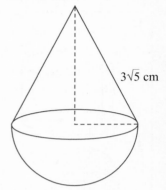

13. (i) The diameter of a solid metal sphere is 9 cm. Find the volume of the sphere in terms of π.
    (ii) The sphere is melted down. All of the metal is used to make a solid shape which consists of a cone on top of a cylinder, as shown in the diagram. The cone and the cylinder both have a height of 8 cm. The cylinder and the base of the cone both have a radius of r cm. Calculate r, correct to one decimal place.

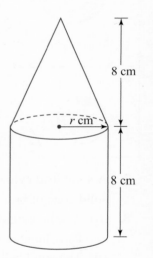

14. (i) A steelworks buys steel in the form of solid cylindrical rods of radius 10 cm and length 30 m. The steel rods are melted to produce solid spherical ball bearings. No steel is wasted in the process. Find the volume of steel in one cylindrical rod in terms of π.
    (ii) The radius of a ball bearing is 2 cm. How many such ball bearings are made from one steel rod?
    (iii) Ball bearings of a different size are also produced. One steel rod makes 225,000 of these new ball bearings. Find the radius of the new ball bearings.

15. (i) A rectangular tank has a height of 4 m and a width of 3 m. Find its length if its volume is 96 m³.
    (ii) The tank is filled with water and a hole is drilled through its base. The water escapes at the rate of 50 litres per second. How long will it take to empty the tank in minutes?
16. Water flows through a cylindrical pipe at the rate of 12 cm per second. The diameter of the pipe is 8 cm. The water flows into an empty cylindrical tank of radius 24 cm. What is the depth of the water after one minute?

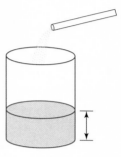

17. (i) A golf ball has a diameter of 3 cm. Find its volume in terms of π.
    (ii) Four golf balls fit exactly into a cylindrical tube, as shown.
        (a) Find the radius and height of the tube.
        (b) Find the volume of the tube in terms of π.
        (c) Find the fraction of the volume of the cylinder that is taken up by the four golf balls.

18. Sweets made from a chocolate mixture are in the shape of solid spherical balls. The diameter of each sweet is 3 cm. 36 sweets fit exactly in a rectangular box which has an internal height of 3 cm.
    (i) The base of the box is a square. How many sweets are there in each row?
    (ii) What is the internal volume of the box?
    (iii) The 36 sweets weigh 675 grams. What is the weight of 1 cm³ of the chocolate mixture? Give your answer correct to one decimal place.
19. A vitamin capsule is in the shape of a cylinder with hemispherical ends. The length of the capsule is 20 mm and the diameter is 6 mm.
    (i) Calculate the volume of the capsule, giving your answer correct to the nearest mm³, taking $\pi = 3 \cdot 14$.

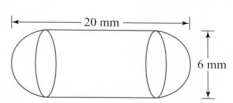

(ii) A course of these vitamins consists of 24 capsules. The capsules are stacked in three rows of eight in a box, as shown in the diagram. Write down:

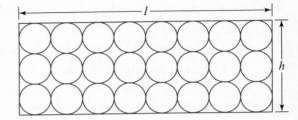

(a) The length, l
(b) The height, h
(c) The width of the box

(iii) Hence, calculate the volume of the box in mm^3.

(iv) How much of the internal volume of the box is not occupied by the capsules?

20. A wax candle is in the shape of a right circular cone. The height of the candle is 7 cm and the diameter of the base is 6 cm.

(i) Find the volume of the wax candle, correct to the nearest cm^3.

(ii) A rectangular block of wax measuring 25 cm by 12 cm by 12 cm is melted down and used to make a number of these candles. Find the maximum number of candles that can be made from the block of wax if 4% of the wax is lost in the process.

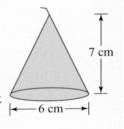

21. (i) Soup is contained in a cylindrical saucepan which has an internal radius of 14 cm. The depth of the soup is 20 cm. Calculate, in terms of π, the volume of soup in the saucepan.

(ii) A ladle in the shape of a hemisphere with an internal radius of length 6 cm is used to serve the soup. Calculate, in terms of π, the volume of soup contained in one full ladle.

(iii) The soup is served into cylindrical cups, each with an internal radius of length 4 cm. One ladleful is placed in each cup. Calculate the depth of the soup in each cup.

(iv) How many cups can be filled from the contents of the saucepan if each cup must contain exactly one full ladle?

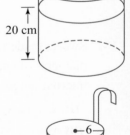

22. The diagram below is a scale drawing of a hopper tank used to store grain. An estimate is needed of the capacity (volume) of the tank. The figure of the man standing beside the tank allows the scale of the drawing to be estimated.

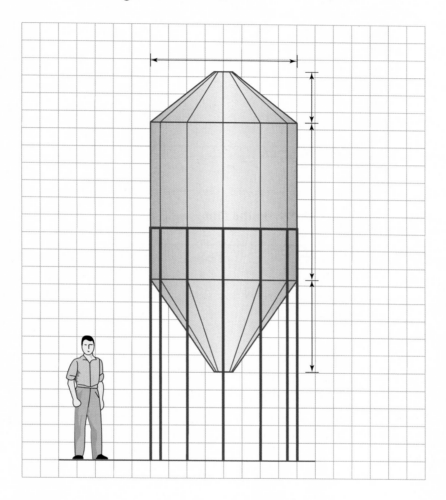

(i) Give an estimate, in metres, of the height of an average adult man.

(ii) Using your answer to part (i), estimate the dimensions of the hopper tank. Write your answers in the spaces provided on the diagram in metres.

(iii) Taking the tank to be a cylinder with a cone above and below, find an estimate for the capacity of the tank in cubic metres.

Trapezoidal rule

The trapezoidal rule gives a concise formula to enable us to make a good approximation of the area of an irregular shape.

Consider the diagram below.

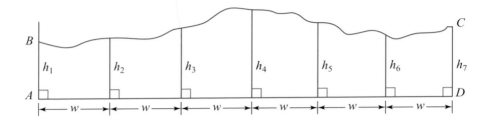

To find the area of the figure *ABCD*, do the following.

1. Divide the figure into a number of strips of equal width. Note: The number of strips can be even or odd.
2. Number and measure each height, h.
3. Use the following formula:
$$\text{Area} = \frac{W}{2}[h_1 + h_7 + 2(h_2 + h_3 + h_4 + h_5 + h_6)]$$
$$\text{Area} = \frac{\text{Width}}{2}[\text{first height} + \text{last height} + 2(\text{sum of all remaining heights})]$$

Note: The greater the number of strips taken, the greater the accuracy.

EXAMPLE 1

Use the trapezoidal rule to estimate the area of the figure below.

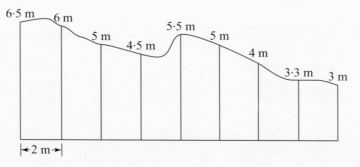

Solution:

$$\text{Area} = \frac{W}{2}[\text{first height} + \text{last height} + 2(\text{sum of all remaining heights})]$$

$$\text{Area} = \frac{W}{2}[h_1 + h_9 + 2(h_2 + h_3 + h_4 + h_5 + h_6 + h_7 + h_8)]$$

Now $h_1 = 6.5$, $h_2 = 6$, $h_3 = 5$, $h_4 = 4.5$, $h_5 = 5.5$, $h_6 = 5$, $h_7 = 4$, $h_8 = 3.3$, $h_9 = 3$

$$\text{Area} = \frac{2}{2}[6.5 + 3 + 2(6 + 5 + 4.5 + 5.5 + 5 + 4 + 3.3)]$$

$$\text{Area} = 1[9.5 + 2(33.3)]$$

$$\text{Area} = 76.1 \text{ m}^2$$

If an irregular shape has no straight edge it can be broken up into two regions, each with its own straight edge, as in the diagram. We then apply the trapezoidal rule in the normal way, except we treat both heights on each side of the line as one height in using the formula (see the next example).

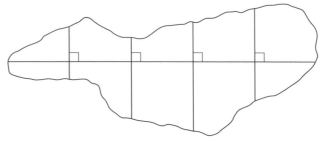

Sometimes we also have to deal with an equation in disguise.

EXAMPLE 2

A surveyor makes the following sketch in estimating the area of a building site, where k is the length shown. Using the trapezoidal rule, she estimates that the area of the site is 175 m². Find k.

Solution:
Estimated area of building site = 175 m².

$$\therefore \frac{W}{2}[h_1 + h_4 + 2(h_2 + h_3)] = 175 \text{ m}$$

$h_1 = 0$
$h_2 = 3 + 8 = 11$
$h_3 = 3 + k$
$h_4 = 0$

$$\frac{7}{2}[0 + 0 + 2(11 + 3 + k)] = 175$$

$$\frac{7}{2}[22 + 6 + 2k] = 175$$

$$77 + 21 + 7k = 175$$

$$7k = 175 - 77 - 21$$

$$7k = 77$$

$$k = 11 \text{ m}$$

Exercise 4.10

Use the trapezoidal rule to estimate the area of the following figures in questions 1–6 (all dimensions are in m).

1.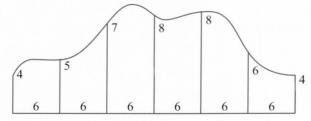

PERIMETER, AREA AND VOLUME

2.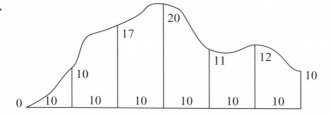

3.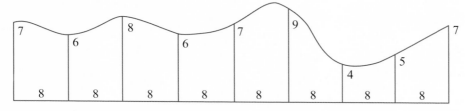

4.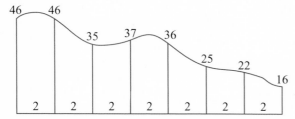

5.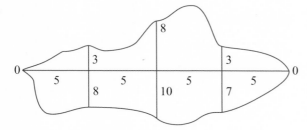

6.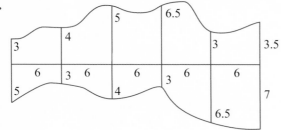

7. A sketch of a piece of land is shown. Using the trapezoidal rule, the area of the piece of land is estimated to be 270 m². Calculate the value of k.

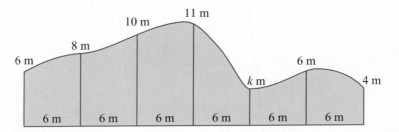

8. Using the trapezoidal rule, the area of the figure below was estimated to be 205 cm². Calculate the value of h.

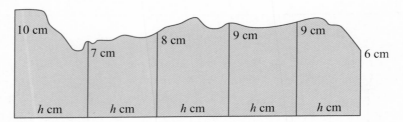

9. The sketch shows a field $ABCD$ which has one uneven edge. At equal intervals of 6 m along $[BC]$, perpendicular measurements of 7 m, 8 m, 10 m, 11 m, 13 m, 15 m and x m are made to the top of the field.

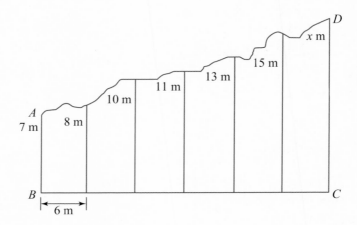

Using the trapezoidal rule, the area of the field is calculated to be 410 m². Calculate the value of x.

10. The sketch shows a flood caused by a leaking underground pipe that runs from A to B.

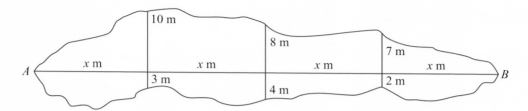

At equal intervals of x m along [AB], perpendicular measurements are made to the edges of the flood. The measurements to the top edge are 10 m, 8 m and 7 m. The measurements to the bottom edge are 3 m, 4 m and 2 m. At A and B, the measurements are 0 m.

Using the trapezoidal rule, the area of the flood is estimated to be 672 m². Find x and hence write down the length of the pipe.

11. Archaeologists excavating a rectangular plot ABCD measuring 120 m by 60 m divided the plot into eight square sections, as shown on the diagram. At the end of the first phase of the work, the shaded area had been excavated. To estimate the area excavated, perpendicular measurements were made to the edge of the excavated area, as shown.

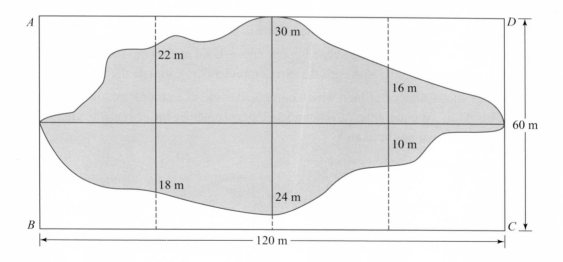

(i) Use the trapezoidal rule to estimate the area excavated.
(ii) Express the excavated area as a percentage of the total area.

12. In order to estimate the area of the irregular shape below, a horizontal line is drawn across the widest part of the shape and three offsets (perpendicular line) are drawn at equal intervals along this line.

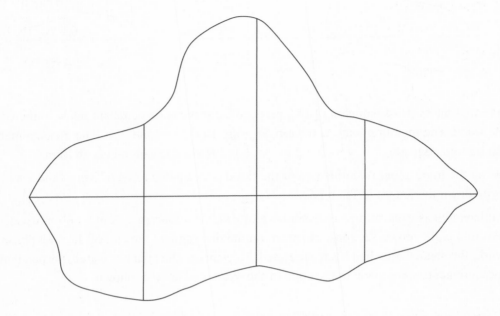

   (i) Measure the horizontal line and the offsets in centimetres.
   (ii) Make a rough sketch of the shape in your answerbook and record the measurements on it. Use the trapezoidal rule with these measurements to estimate the area of the shape.

13. The sketch shows a piece of land which borders the side of a straight road [AB]. The length of [AB] is 63 m. At equal intervals along [AB], perpendicular measurements are made to the boundary, as shown on the sketch.

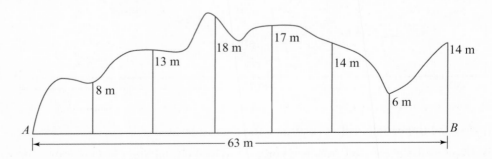

   (i) Use the trapezoidal rule to estimate the area of the piece of land.
   (ii) The land is valued at €380,000 per hectare. Find the value of the piece of land.
      (**Note:** 1 hectare = 10,000 m^2)

PERIMETER, AREA AND VOLUME

14. The diagram shows a quadrant of a circle of radius 10.

    (i) Using the trapezoidal rule, estimate the area of the quadrant (all dimensions are in cm).

    (ii) Using the formula for the area of a circle, πr^2, find the area of the quadrant with $\pi = 3 \cdot 14$.

    (iii) Find the error between (i) and (ii).

    (iv) Which answer do you think is most accurate? Justify your answer.

    (v) Hence, calculate the percentage error correct to the nearest integer.

    (vi) The diagram shows the same quadrant divided into five strips of equal width. Use the trapezoidal rule to estimate the area of the quadrant.

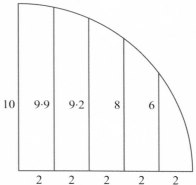

    (vii) Given the same quadrant divided into eight strips of equal width, write down your estimate for the area without doing any calculations. Justify your answer.

15. (i) The front face of a stone wall of a ruined castle is shown in the diagram. All distances are measured in metres. The heights are measured at intervals of 2·4 m along the base line. Use the trapezoidal rule to calculate the area of the front face of the stone wall.

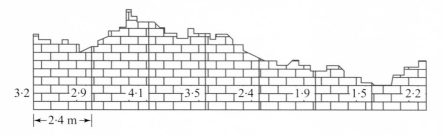

    (ii) An archaeologist estimates that this ruined wall is 16% of the original wall. Find the area of the original wall.

16. The depth of the water in a river of width 20 m was measured at intervals of 4 m, starting from one bank and ending at the other. The results are recorded in the following table.

Distance from the bank (m)	0	4	8	12	16	20
Depth (m)	0·6	0·8	1·4	2·1	1·9	0·7

(i) Use the trapezoidal rule to estimate the area of a cross-section of the water at this point.

(ii) At this point of the river, it was calculated that the water was flowing at a speed of $\frac{1}{2}$ m per second. Find the volume of water passing this point each minute **(a)** in cubic metres **(b)** in litres.

17. The diagram shows the curve $y = x^2 + 2$ in the domain $0 \leq x \leq 5$, where $x \in \mathbb{R}$.

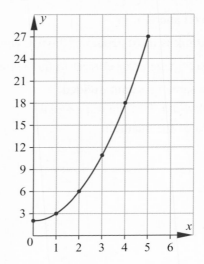

(i) Copy the following table, then complete it using the equation of the curve.

x	0	1	2	3	4	5
y						

(ii) Hence, use the trapezoidal rule to estimate the area between the curve and the x-axis.

(iii) The actual area between the curve and the x-axis is $51\frac{2}{3}$ square units. Find the percentage error in the estimate, correct to one decimal place.

(iv) Comment on how effective you think the trapezoidal rule is in this case.

18. A Geography class is required to estimate the area of Carlingford Lough from a map with a scale of 1 : 70,000.

An axis line *ABCDEF* is drawn and divided into five equal segments (of 3 cm each), as shown.

(i) Complete the following table by writing down the values of x and y by measuring from the map. Give your answers correct to the nearest half centimetre.

Through the point	A	B	C	D	E	F
Perpendicular distance from shore to shore in cm	0	$3\frac{1}{2}$	$3\frac{1}{2}$	x	y	$2\frac{1}{2}$

PERIMETER, AREA AND VOLUME

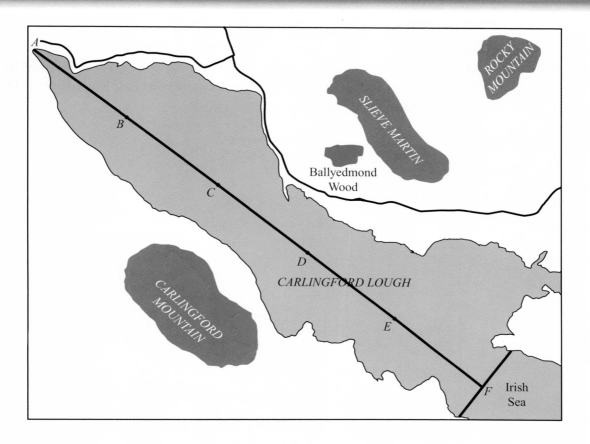

(ii) Use the trapezoidal rule to estimate the following.

(a) The area in cm² of Carlingford Lough as outlined on the map, correct to the nearest cm².

(b) The area in km² of Carlingford Lough as outlined, correct to the nearest km².

(c) Comment on the accuracy or otherwise of the trapezoidal rule in this situation. Justify your comment.

19. The diagram below shows a shape with two straight edges and one irregular edge. By dividing the edge [AB] into five equal intervals, use the trapezoidal rule to estimate the area of the shape.
Record your constructions and measurements on the diagram. Give your lengths correct to the nearest half cm and your area correct to the nearest cm².

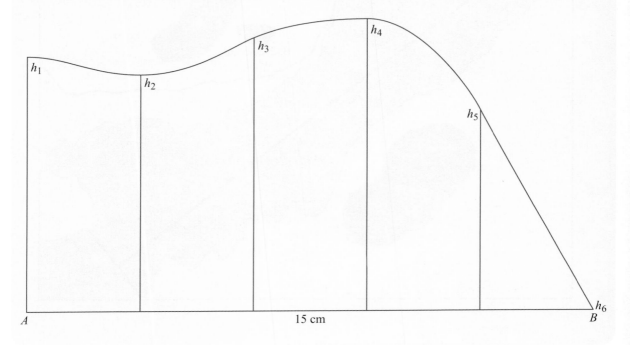

Nets of 3D shapes

A line has only one **dimension** – length (1D).

A flat shape has two dimensions – length and width (2D).

A **solid** shape has three dimensions – length, width and height (3D).

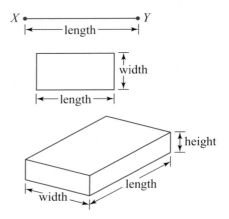

PERIMETER, AREA AND VOLUME

When a 3D shape is opened out, the flat shape is called the **net**.

 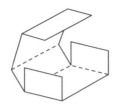 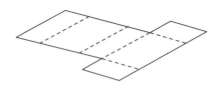

EXAMPLE

(i) This is a **net** of a solid cube.

This is how it folds up to make the cube.

There can be many different nets for one rectangular solid, e.g. this is also a net of a solid cube.

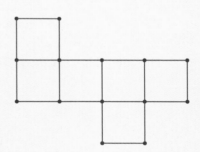

(ii) This is a net for a cuboid that is 4 cm by 2 cm by 2 cm.
When you draw a net, you have to draw the lengths accurately.
You may have to use a scale for your drawing.
Choose a scale so that the net fits on your page and it's not too small.

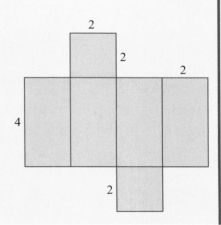

This is how the net folds up to make a cuboid.

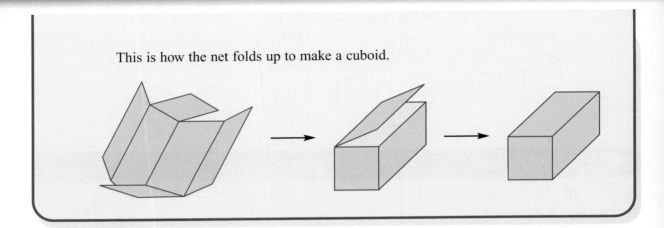

Naming parts of a 3D shape

Each flat surface is called a **face**. Two faces meet at an **edge**. Edges of a shape meet at a corner, or point, called a **vertex**. The plural of vertex is **vertices**.

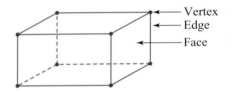

The cuboid has eight vertices (each marked with a •).
The cuboid has 12 edges (count each line, including dotted lines).
The cuboid has six faces. They are front and back (not indicated) plus top, base and sides.

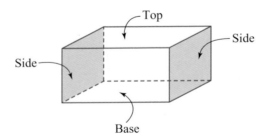

A net diagram indicates the faces clearly.

	Top	
	Back	
Side	Base	Side
	Front	

PERIMETER, AREA AND VOLUME

EXAMPLE

Using a net to determine the surface area of a 3D shape, find the surface area of the cuboid with dimensions height 7 cm, length 10 cm and width 6 cm.

Solution:

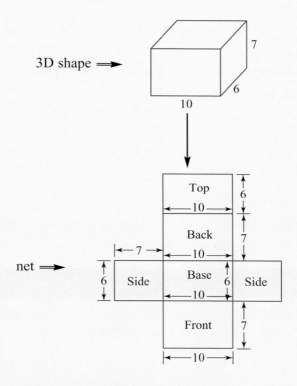

Surface area of a cuboid = The sum of the area of all six faces
  = Sum of the area of all six faces of the net
  = Area of (top + back + base + front + side + side)
  = Area of ((top + base) + (back + front) + (side + side))
  = 2(top) + 2(back) + 2(side)
  = 2(10 × 6) + 2(10 × 7) + 2(6 × 7)
  = 120 + 140 + 84
  = 344 cm^2

Note: We can write a formula for the surface area of a cuboid as $2lb + 2lh + 2bh$.

Examples of 3D shapes (or solids)

These are all examples of 3D shapes.

Cube

Triangular prism

Pyramid with square base

Sphere

Can you name some other 3D shapes?

These 3D shapes are called prisms.

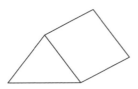

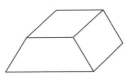

What do prisms have in common?

Can you draw another 3D shape that is a prism?

EXAMPLE

This prism is 8 cm long.
The ends are equilateral triangles with sides of 4 cm.
Draw an accurate net of the prism.

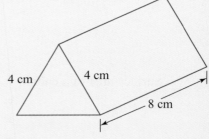

Solution:

Step 1: Draw the rectangular faces of the prism.
Each rectangle is 8 cm long and 4 cm wide.

Step 2: The ends of the prism are equilateral triangles.
The length of each side of the triangles is 4 cm.
Use your compass to construct the equilateral triangles.

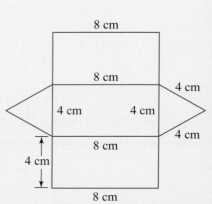

PERIMETER, AREA AND VOLUME

Exercise 4.11

1.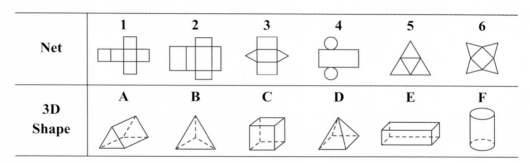

   The table shows six nets numbered 1, 2, 3, 4, 5, 6 and six 3D shapes labelled A, B, C, D, E, F. Match each net to its correct 3D shape. The first one is done for you: (1, C).

2. Sam made this shape using multilink cubes. Four cubes are purple. The other six are white.

   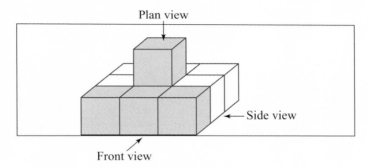

   Copy and complete each view of the shape by colouring in the squares that are purple.

   (i)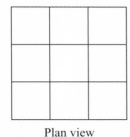

   Plan view

   (ii)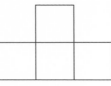

   Front view

   (iii)

   Side view

3. Copy each diagram and colour in two extra squares to give you the net of a cube.

   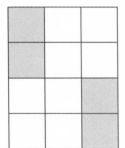 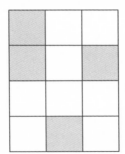 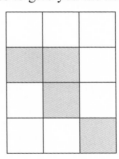

4. Draw an accurate net for each of these 3D shapes.

   (i) (ii) (iii)

   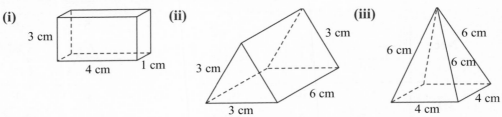

5. Here is the net of a triangular prism with dimensions in cm.

   (i) State the number of faces for this prism.
   (ii) How many vertices has the prism?
   (iii) Find (a) the surface area of the prism (b) the volume of the prism.

   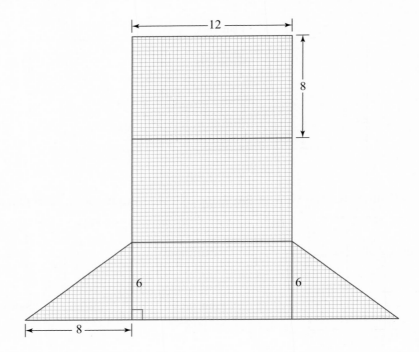

PERIMETER, AREA AND VOLUME

6. The nets of these shapes have reflection symmetry. Complete the nets and name the two shapes.

(i)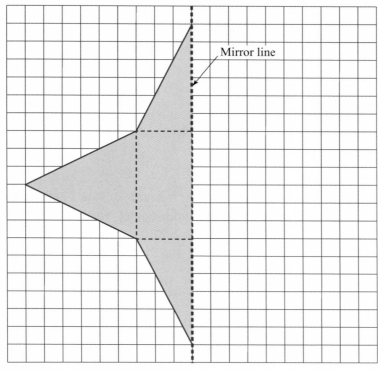

(ii)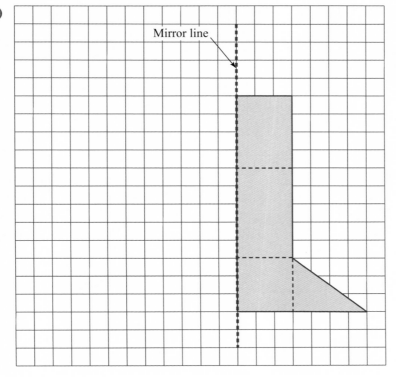

7. (i) A metal box with no lid is shown. Draw a net of the box.
    (ii) Hence or otherwise, find the area of the metal required to make this box. Give your answers in
        (a) cm² (b) m².

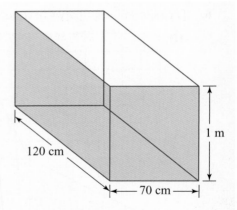

8. A rectangular sheet of metal measures 2 m by $1\frac{1}{2}$ m. A square of side 30 cm is removed from each corner. The remaining piece of metal is folded along the dotted lines to form an open box as shown.
    (i) Find the surface area of the net used to construct the box.
    (ii) Find the volume of the box in (a) cm³ (b) m³ (c) litres.

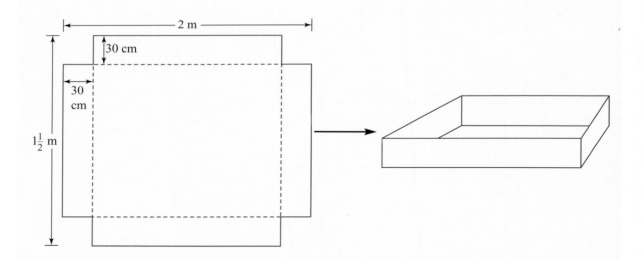

9. Look at these diagrams of 3D shapes. Dotted lines are used to show the edges that cannot be seen when you look at the shape from one side.

(i)   (ii) 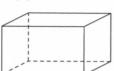 (iii)   (iv)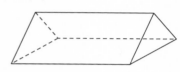

Copy and complete this table.

	Name of shape	Number of faces	Number of vertices	Number of edges
(i)				
(ii)				
(iii)				
(iv)				

10. (i) Which 3D solids could be made from the following nets?

(a) 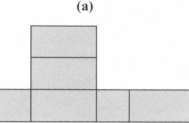 (b)   (c)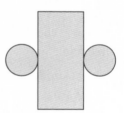

(ii) Find the total surface area of the nets of the solids below. Take $\pi = \dfrac{22}{7}$. (All dimensions are in cm.)

(a) 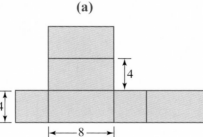 (b) 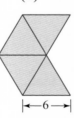 (c)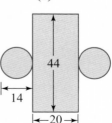

(iii) Find the volume of the solids in (a) and (c), taking $\pi = \dfrac{22}{7}$.

11. The following net has a 3D shape with a volume of 2,512 cm³.

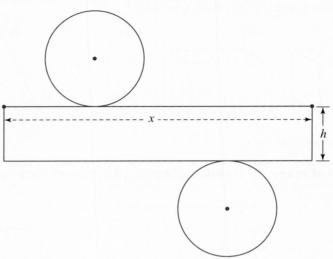

(i) Name the 3D shape.

(ii) Taking $\pi = 3{\cdot}14$ and given both identical circles have a diameter of 20 cm, calculate h and x, the dimensions of the rectangle.

(iii) Can a similar 3D solid be made from the following net? Justify your answer.

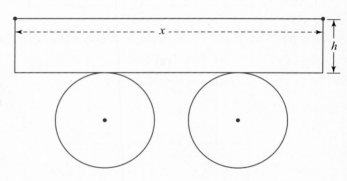

12. Find the volume of the prism associated with the given net. (All dimensions are in cm.)

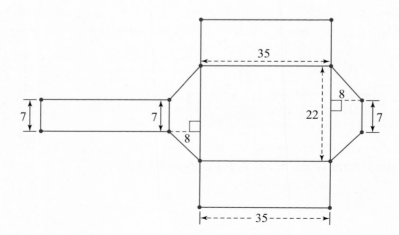

CHAPTER 5 PATTERNS, SEQUENCES AND SERIES

Patterns 1

Much of mathematics is about patterns. Some are simple and numeric:

$$2, 4, 6, 8, 10, \ldots \qquad \frac{1}{2}, \frac{2}{3}, \frac{3}{4}, \frac{4}{5}, \ldots \qquad 10, 21, 32, 43, 54, \ldots$$

You should be able to write down the next three numbers in the list; you may even be able to predict the 10th number in the list *without* having to write out all the terms.

Other numeric patterns are more complicated:

$$1, 2, 4, 8, 16, \ldots \qquad 1, 0, 2, 0, 0, 3, 0, 0, 0, 4, \ldots \qquad 1, 2, 6, 24, 120, 720, \ldots$$

Even if you see a pattern and can write out the next number, it is much more difficult to predict what the 20th number or the 100th number in the list would be. If those numbers represented the population of the planet or the number of cancerous cells in a patient, then it would be very important to be able to predict future values.

Not all patterns are numeric. For example:

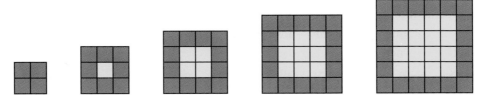

Is it possible to predict the number of squares in the 10th diagram? What about the number of yellow squares in the 10th diagram? What about the purple squares?

Many young children like to watch how tall they are growing and use some simple measuring techniques to record their growth. Is it possible to predict a child's height as each year passes?

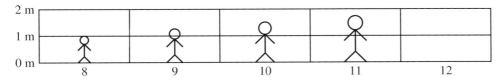

As an eight-year old, Freddie was 1 m tall. By nine he was 1·25 m tall and by 10 he had reached 1·5 m. Could you use this information to calculate his height when he is 12 years old? What about when Freddie is 21? Do you see a problem?

Sequences

> A **sequence** is a set of numbers, separated by commas, in which each number after the first is formed by some definite rule.

> Each number in a sequence is a **term** of that sequence. The first number is the **first term** and is denoted by T_1. Similarly, the second term is denoted by T_2 and so on.

3, 7, 11, 15, . . .
Each number after the first is obtained by adding 4 to the previous number. In this example, 3 is called the **first term**, 7 is the **second term** and so on.
1, 3, 9, 27, . . .
Each number after the first is obtained by multiplying the previous number by 3. In this example, 1 is called the **first term**, 3 is the **second term** and so on.

The general term, T_n

Very often a sequence is given by a **rule** which defines the **general term**. We use T_n to denote the general term of the sequence. T_n may be used to obtain any term of a sequence. T_1 will represent the first term, T_2 the second term and so on.

Notes: 1. The general term, T_n, is often called the *n*th term.
2. *n* used with this meaning must always be a positive whole number. It can never be fractional or negative.
3. A sequence is often called a progression.

Consider the sequence whose general term is $T_n = 3n + 2$.

We can find the value of any term of the sequence by putting in the appropriate value for *n* on both sides:

$$T_n = 3n + 2$$

$T_1 = 3(1) + 2 = 3 + 2 = 5$ (first term, put in 1 for *n*)

$T_2 = 3(2) + 2 = 6 + 2 = 8$ (second term, put in 2 for *n*)

$T_5 = 3(5) + 2 = 15 + 2 = 17$ (fifth term, put in 5 for *n*)

In each case, *n* is replaced with the same number on both sides.

The notation $T_n = 3n + 2$ is very similar to function notation when *n* is the input and T_n is the output, i.e. (input, output) = (n, T_n).

EXAMPLE

The nth term of a sequence is given by $T_n = n^2 + 3$.

(i) Write down the first three terms of the sequence.

(ii) Show that: (a) $\dfrac{T_5}{T_2} = T_1$ (b) $2T_4 = T_6 - 1$

Solution:

(i) $T_n = n^2 + 3$

$T_1 = 1^2 + 3 = 1 + 3 = 4$ (put in 1 for n)

$T_2 = 2^2 + 3 = 4 + 3 = 7$ (put in 2 for n)

$T_3 = 3^2 + 3 = 9 + 3 = 12$ (put in 3 for n)

Thus, the first three terms are 4, 7, 12.

(ii) (a) From (i), $T_1 = 4$ and $T_2 = 7$.

$T_5 = 5^2 + 3 = 25 + 3 = 28$

$\dfrac{T_5}{T_2} = \dfrac{28}{7} = 4$

$T_1 = 4$

$\therefore \dfrac{T_5}{T_2} = T_1$

(b) $T_4 = 4^2 + 3 = 16 + 3 = 19$

$T_6 = 6^2 + 3 = 36 + 3 = 39$

$2T_4 = 2(19) = 38$

$T_6 - 1 = 39 - 1 = 38$

$\therefore 2T_4 = T_6 - 1$

Exercise 5.1

In questions 1–8, write down the next four terms.

1. 1, 5, 9, 13, . . .
2. 40, 35, 30, 25, . . .
3. −11, −9, −7, −5, . . .
4. 13, 10, 7, 4, . . .
5. 2·5, 2·9, 3·3, 3·7, . . .
6. 2·8, 2·2, 1·6, 1, . . .
7. 1, 2, 4, 8, . . .
8. 2, 6, 18, 54, . . .

In questions 9–20, write down the first four terms of the sequence defined by the given nth term.

9. $T_n = 2n + 3$
10. $T_n = 3n + 1$
11. $T_n = 4n - 1$
12. $T_n = 5n - 3$
13. $T_n = 1 - 2n$
14. $T_n = 3 - 4n$
15. $T_n = n^2 + 5$
16. $T_n = n^2 + 2n$
17. $T_n = \dfrac{n + 1}{n}$
18. $T_n = \dfrac{2n}{n + 1}$
19. $T_n = 2^n$
20. $T_n = 3^n$

21. The nth term of a sequence is given by $T_n = 5n + 2$.
    (i) Write down the first three terms of the sequence.
    (ii) Show that: (a) $2T_5 = T_4 + T_6$ (b) $6(T_7 - 1) = T_2(T_3 + 1)$

22. The nth term of a sequence is given by $T_n = n^2 + 2$.
    (i) Write down the first three terms of the sequence.
    (ii) Show that: (a) $\dfrac{T_4}{T_2} = T_1$ (b) $\dfrac{T_6 - 2}{T_4} = \dfrac{T_2}{T_1}$

23. The nth term of a sequence is given by $T_n = \dfrac{n+2}{n+1}$.
    (i) Write down T_1, T_2 and T_3, the first, second and third terms.
    (ii) Show that $T_1 + T_2 > 2T_3$.

Patterns 2

A pattern of repeating coloured blocks is easy enough to continue, but the challenge is to be able to predict what happens much further along.

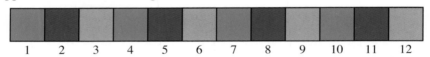

Looking at the orange blocks, there is an interesting pattern:

The 1st orange block is at position 3	$3 = 3 \times 1$
The 2nd orange block is at position 6	$6 = 3 \times 2$
The 3rd orange block is at position 9	$9 = 3 \times 3$
The 4th orange block is at position 12	$12 = 3 \times 4$

The 8th orange block should be at position $3 \times 8 = 24$.

We can predict that there will be another orange block at position $3 \times 9 = 27$. We can jump much further along and predict an orange block at position 99 (because $3 \times 33 = 99$).

So what about the 100th block? If the 99th is orange, then the 100th must be green and the 101st must be blue. To understand the sequence of this pattern, we need to find one sequence of colours which is easy to predict. From that we can deduce the positions of the other colours.

EXAMPLE

A repeating pattern consists of blocks coloured yellow, purple, green, yellow, . . . and so on.

(i) Complete the table.

Block position	Colour
1	Yellow
2	
	Green
⋮	⋮

(ii) List the positions of the first three yellow blocks. Is there a pattern?
(iii) List the positions of the first three purple blocks. Is there a pattern?
(iv) List the positions of the first three green blocks. Is there a pattern?
(v) What is the colour of the 48th block?
(vi) What is the colour of the 50th block?
(vii) What is the colour of the 100th block?

Solution:

(i) The completed table:

Block position	Colour
1	Yellow
2	Purple
3	Green
4	Yellow
5	Purple
6	Green
7	Yellow
8	Purple
9	Green

(ii) Yellow blocks: 1, 4, 7, . . .
  Yes. Starting with 1 and adding 3 each time, the other positions can be found.

(iii) Purple blocks: 2, 5, 8, . . .
  Yes. Starting with 2 and adding 3 each time, the other positions can be found.

(iv) Green blocks: 3, 6, 9, . . .
  Yes. Starting with 3 and adding 3 each time, the other positions can be found.
  Alternatively, the positions are multiples of 3, so the 1st green is at position 3 × 1 = 3, the 2nd green is at position 3 × 2 = 6, the 3rd green is at position 3 × 3 = 9 and so on.

(v) As 48 is a multiple of 3 (48 can be divided exactly by 3), it must be a green block.

(vi) Since the 48th block is green, the 50th must be purple.

(vii) Using a calculator, 100 is not a multiple of 3 (it divides in just over 33 times). $3 \times 33 = 99$ which *is* a multiple of 3. So the 99th block must be green.

The 100th block is therefore yellow.

Exercise 5.2

1. A repeating pattern consists of blocks coloured green, yellow, green, yellow, . . . and so on.

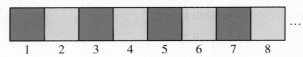

   (i) Complete the table.

Block position	Colour
1	Green
2	Yellow
⋮	⋮

   (ii) List the positions of the first three green blocks. Is there a pattern?
   (iii) List the positions of the first three yellow blocks. Is there a pattern?
   (iv) What is the colour of the 20th block?
   (v) What is the colour of the 33rd block?
   (vi) What is the colour of the 1,001st block?

2. A repeating pattern consists of blocks coloured red, white, blue, red, . . . and so on.

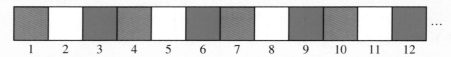

   (i) List the positions of the first three red blocks. Is there a pattern?
   (ii) List the positions of the first three white blocks. Is there a pattern?
   (iii) List the positions of the first three blue blocks. Is there a pattern?
   (iv) What is the colour of the 48th block?
   (v) What is the colour of the 50th block?
   (vi) What is the colour of the 100th block?

3. Players in a football competition are lined up in the following repeating sequence.

    (i) List the positions of the first four players in a blue shirt.
    (ii) List the positions of the first four players in an orange shirt.
    (iii) List the positions of the first four players in a yellow shirt.
    (iv) What is the colour of the shirt of the 15th player?
    (v) What is the colour of the shirt of the 20th player?
    (vi) Where in the line-up is the 8th player with an orange shirt?
    (vii) Where in the line-up is the 10th player with a blue shirt?

4. As part of the opening ceremony of a hockey competition, the hockey players parade in the following repeating sequence.

    (i) List the positions of the first three players in each colour.
    (ii) What is the colour of the shirt of the 12th player?
    (iii) What is the colour of the shirt of the 22nd player?
    (iv) What is the colour of the shirt of the 23rd player?
    (v) Where in the line-up is the 8th player with a green shirt?
    (vi) Where in the line-up is the 10th player with a yellow shirt?

Differences – 1

Some sequences can be understood better by investigating the differences between the terms. For example, the sequence 2, 5, 8, . . . has the same difference between consecutive terms. We can say that the difference between the terms is constant.

The differences are equal to 3 (which is a constant).

More complex sequences require us to check **the difference between the differences**. For example, the sequence 3, 6, 11, 18, 27, . . .

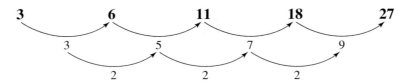

The first differences (3, 5, 7, . . .) are not constant, but the second differences are equal to 2 (which is a constant).

Exercise 5.3

In questions 1–6, find the first four terms and show that the difference between the terms is constant.

1. $T_n = 2n + 1$
2. $T_n = 3n + 2$
3. $T_n = 4n - 3$
4. $T_n = 5n + 2$
5. $T_n = 3 - 2n$
6. $T_n = 8 - 3n$

In questions 7–9, find the first four terms and show that the difference between the differences is constant.

7. $T_n = n^2 + 1$
8. $T_n = n^2 - 2$
9. $T_n = n^2 + n$

In questions 10–12, find the first four terms and show that the differences are such that this method of investigation will not produce a useful result.

10. $T_n = 2^n$
11. $T_n = 3^n$
12. $T_n = 3 + 2^n$

Viewing the sequence as a graph

The patterns we look for are based on the relationship between two sets of values – the positions in a sequence and the values or terms in the sequence.

For example, the sequence

$$1, 3, 5, 7, 9, \ldots$$

can be represented in this table:

Position	1	2	3	4	5
Terms	1	3	5	7	9

We can interpret the pairs (position, term) as points and plot (1, 1), (2, 3), (3, 5), (4, 7) and (5, 9).
The graph shows a line.

Note: The first four letters of the word **linear** spell line. A graph is linear if all the points can be joined by one straight line.

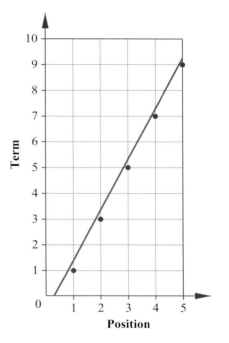

Exercise 5.4

In questions 1–8, find the first five terms and show your results on a graph. State whether the graph is linear or not. Put the position along the horizontal axis.

1. $T_n = 2n$
2. $T_n = 3n - 1$
3. $T_n = 4n + 1$
4. $T_n = n^2$
5. $T_n = n^2 - n$
6. $T_n = n(n + 2)$
7. $T_n = 2^n$
8. $T_n = n^3 - n^2$

9. (i) Find the first four terms in the sequence defined by $T_n = 2n + 3$.
   (ii) By forming pairs of the form (position, term), plot the sequence on a graph.
   (iii) Using $x = 1, 2, 3$ and 4, find 4 points on the line $y = 2x + 3$. What do you notice?

10. (i) Find the first four terms in the sequence defined by $T_n = 3n - 2$.
    (ii) By forming pairs of the form (position, term), plot the sequence on a graph.
    (iii) Using $x = 1, 2, 3$ and 4, find 4 points on the line $y = 3x - 2$. What do you notice?
    (iv) What is the slope of the line in your graph?
    (v) What is the connection between the slope and the numbers in the original sequence?

11. Two sequences are defined as follows:

    Sequence A: $T_n = 2n + 7$ Sequence B: $T_n = 7n + 2$

    (i) In which sequence will the terms increase more rapidly? Explain your answer.
    (ii) Verify your result by finding T_1 and T_5 of each sequence and compare your results.
    (iii) Use your results from part (ii) to sketch a graph showing the two sequences as lines.

12. John receives a gift of a money box with €4 in it for his birthday. John decides he will save a further €2 a day each day after his birthday.
   (i) Draw a table showing the amount of money John saved for the first five days (his birthday being the first day).
   (ii) How much money will John have in his money box on the 10th day?
   (iii) How much money will John have in his money box on the 25th day?
   (iv) By looking at the pattern in this question, can you explain why the amount of money John has on day 10 is not twice the amount he has on day 5?
   (v) How much money does John have in his money box on day 100?
   (vi) How much money has John actually put in his money box after 10 days? Explain how you arrived at this amount.
   (vii) John wants to buy a new computer game. The game costs €39.99. What is the minimum number of days John will have to save so that he has enough money to buy the computer game?

13. Owen has a money box. He starts with €1 and adds €3 each day.
   (i) Draw a table showing the amount of money Owen has for the first 10 days.
   (ii) Draw a graph to show the amount of money Owen has saved each day. Put *Number of days* along the horizontal axis and *Amount of money* on the vertical axis.
   (iii) Why will the scale for the *Number of days* be different to the scale for the *Amount of money*?

14. Amy and Bill are discussing phone network offers. Bill says that on his network he begins each month with 30 free texts and receives three additional free texts each night. Amy says that she gets no free texts at the beginning of the month but that she receives five free texts each night.
   (i) Draw a table showing the start number of texts and the number of texts available for the first six days.

Day	Bill's texts	Amy's texts
Start (1)	0	30
Day 2	5	33
⋮		
Day 6		

   (ii) Who has the most free texts after six days?
   (iii) Using graph paper, draw a graph showing the number of texts for both Bill and Amy for the first 20 days. You should allow the vertical axis to reach 150 texts.

(iv) Are the two graphs linear or something else?

(v) At what point does Amy seem to have a better offer than Bill?

(vi) Will Amy and Bill ever have the same number of texts on a particular day? If so, which day? If not, why not?

15. Lenny begins the month with 20 free texts and receives two additional free texts each night. Jane does not have any free texts at the beginning of the month but receives three free texts each night.

    (i) Draw a table showing the number of free texts for both Lenny and Jane for the first 10 days.

    (ii) Represent both sets of results on a graph.

    (iii) Will Lenny and Jane ever have the same number of free texts on a certain day? If so, which day? If not, why not?

    (iv) Who in your opinion has the better deal for free texts each month? Give a reason for your answer.

16. A yellow flower measured 5 cm high at the beginning of the week and grew 2 cm each week afterwards.

    A red flower measured 8 cm high at the beginning of the same week but grew only 1·5 cm each week afterwards.

    (i) Calculate the heights of each plant for the first five weeks and plot your results on a graph.

    (ii) Which plant will be taller?

    (iii) Will the plants ever be the exact same height?

    (iv) Give some reasons why the growth suggested by the graph may be unreliable.

Arithmetic sequence 1

Consider the sequence of numbers 2, 5, 8, 11, . . .

Each term after the first can be found by adding 3 to the previous term.

This is an example of an arithmetic sequence.

> A sequence in which each term after the first is found by adding a constant number is called an **arithmetic sequence**.

The first term of an arithmetic sequence is denoted by a. In other words, $T_1 = a$.

The constant number, which is added to each term, is called the **common difference** and is denoted by d.

Consider the arithmetic sequence 3, 5, 7, 9, 11, . . .

$$a = 3 \text{ and } d = 2$$

Each term after the first is found by adding 2 to the previous term.

Consider the arithmetic sequence 7, 2, −3, −8, . . .

$$a = 1 \text{ and } d = -5$$

Each term after the first is found by subtracting 5 from the previous term.

In an arithmetic sequence, the difference between any two consecutive terms is always the same.

$$\boxed{\text{Any term} - \text{previous term} = T_n - T_{n-1} = \text{constant} = d}$$

General term of an arithmetic sequence

In an arithmetic sequence, a is the first term and d is the common difference.
Thus, in an arithmetic sequence:

$T_1 = a$
$T_2 = a + d$
$T_3 = a + 2d$
$T_4 = a + 3d$, etc.

Notice the coefficient of d is always one less than the term number.

Examples:
$T_{10} = a + 9d$
$T_{15} = a + 14d$
$T_{50} = a + 49d$

To go from one term, just add on another d.

$$\boxed{T_n = a + (n-1)d}$$

Note: Once we find a and d, we can answer any question about an arithmetic sequence.

EXAMPLE 1

The first three terms of an arithmetic sequence are 5, 8, 11, . . .

(i) Find the first term, a, and the common difference, d.

(ii) Find, in terms of n, an expression for T_n, the nth term, and hence or otherwise, find T_{17}.

(iii) Which term of the sequence is 122?

Solution:

(i) Find a and d. The first three terms are 5, 8, 11, . . .

$$a = T_1 = 5 \qquad d = T_2 - T_1 = 8 - 5 = 3$$

Note: Be careful: $d \neq T_1 - T_2$.

(ii) Find an expression for T_n.

$$\begin{aligned}T_n &= a + (n-1)d \\ &= 5 + (n-1)3 \\ &= 5 + 3n - 3 \\ &= 3n + 2\end{aligned}$$

Find T_{17}.

Method 1:

$$T_n = a + (n-1)d$$
$$T_{17} = 5 + (17-1)(3)$$

(put in $n = 17$, $a = 5$ and $d = 3$)

$$T_{17} = 5 + 16(3)$$
$$T_{17} = 53$$

Method 2:

$$T_n = 3n + 2 \quad \text{(from part (ii))}$$
$$T_{17} = 3(17) + 2$$

(put in $n = 17$)

$$T_{17} = 53$$

(iii) Which term of the sequence is 122?

Method 1:

Equation given in disguise

$$T_n = 122$$
$$a + (n-1)d = 122$$

(we know a and d, find n)

$$5 + (n-1)(3) = 122$$
$$5 + 3n - 3 = 122$$
$$3n + 2 = 122$$
$$3n = 120$$
$$n = 40$$

Thus, the 40th term is 122.

Method 2:

Equation given in disguise

$$T_n = 122$$

($T_n = 3n + 2$, find n)

$$3n + 2 = 122$$
$$3n = 120$$
$$n = 40$$

EXAMPLE 2

The first three terms in an arithmetic sequence are $k + 2, 2k + 3, 5k - 2$, where k is a real number. Find the value of k and write down the first three terms.

Solution:
We use the fact that in an arithmetic sequence, the difference between any two consecutive terms is always the same. We are given the first three terms.

$$\therefore \quad T_3 - T_2 = T_2 - T_1 \quad \text{(common difference)}$$
$$(5k - 2) - (2k + 3) = (2k + 3) - (k + 2) \quad \text{(put in given values)}$$
$$5k - 2 - 2k - 3 = 2k + 3 - k - 2$$
$$3k - 5 = k + 1$$
$$2k = 6$$
$$k = 3$$

$T_1 = k + 2 = 3 + 2 = 5$
$T_2 = 2k + 3 = 2(3) + 3 = 9$
$T_3 = 5k - 2 = 5(3) - 2 = 13$

Thus, the first three terms are
5, 9, 13.

Exercise 5.5

In questions 1–9, find a and d for the arithmetic sequences and find, in terms of n, an expression for T_n, the nth term.

1. 1, 3, 5, . . .
2. 2, 5, 8, . . .
3. 3, 7, 11, . . .
4. 6, 11, 16, . . .
5. 9, 7, 5, . . .
6. 4, 1, −2, . . .
7. 8, 3, −2, . . .
8. 4, −2, −8, . . .
9. −5, −3, −1, . . .

10. The first three terms of an arithmetic sequence are 1, 4, 7.
    (i) Find the first term, a, and the common difference, d.
    (ii) Find, in terms of n, an expression for T_n, the nth term, and hence or otherwise, find T_{50}.
    (iii) Which term of the sequence is 88?

11. The first three terms of an arithmetic sequence are 4, 9, 14.
    (i) Find the first term, a, and the common difference, d.
    (ii) Find, in terms of n, an expression for T_n, the nth term, and hence or otherwise, find T_{45}.
    (iii) Which term of the sequence is equal to 249?

12. The first three terms of an arithmetic sequence are 40, 36, 32.
    (i) Find the first term, a, and the common difference, d.
    (ii) Find, in terms of n, an expression for T_n, the nth term, and hence or otherwise, find T_{15}.
    (iii) Which term of the sequence is 0?

13. The cost of visiting an exhibition is as follows.

Number of children	Price (€)
1	5
2	7
3	9
4	11

Number of adults	Price (€)
1	7
2	12
3	17
4	22

(i) Find a formula to calculate the cost of n children visiting the exhibition.

(ii) Verify that your formula works for three children.

(iii) Find a formula to calculate the cost of n adults visiting the exhibition.

(iv) Verify that your formula works for four adults.

(v) A group of six adults and 10 children are planning to visit the exhibition. How much will it cost?

14. A taxi charges €5 for a journey of 1 km, €8 for 2 km, €11 for 3 km and so on.

(i) Write out a table showing the cost of journeys up to 6 km.

(ii) Treating the costs as an arithmetic sequence, find the first term, a, and the common difference, d.

(iii) The taxi fare is a fixed charge plus a rate per kilometre. What is (a) the fixed charge and (b) the rate per kilometre?

(iv) How long a journey can be made with €100?

15. The instructions for cooking a chicken are 15 minutes per kg plus 20 minutes.

(i) How long is needed to cook a 1 kg chicken?

(ii) Write out a table showing the time needed to cook chickens weighing 1 kg, 2 kg, . . . , 6 kg.

(iii) Treating the times as an arithmetic sequence, find the first term, a, and the common difference, d.

(iv) Find a formula which calculates the time, in minutes, to cook a chicken weighing n kg.

(v) Use your formula to find out how long will it take to cook a 9 kg chicken.

(vi) What is the heaviest chicken that can be cooked within 4 hours?

16. 5, 8, 11, . . . is an arithmetic sequence. Which term of the sequence is 179?
17. 3, 8, 13, 18, . . . is an arithmetic sequence. Which term of the sequence is 198?
18. Meriel's vocabulary was checked every two months beginning when she was 10 months old. At that time, she had a vocabulary of eight words. At 12 months she had a vocabulary of 11 words. By 14 months it was 14 words and by 16 months it was 17 words.

    (i) Taking the months as a sequence, how old was she when the 10th check was made?

    (ii) Find a formula, in terms of n, which will calculate Meriel's age when the nth check is made.

    (iii) How old was she when the 10th check was made?

    (iv) Having a vocabulary of 5,000 words is not unusual. How old will Meriel be when she knows 5,000 words?

19. The first three terms in an arithmetic sequence are $2k - 1$, $2k + 1$, $3k$, where k is a real number. Find the value of k and write down the first three terms.

20. The first three terms in an arithmetic sequence are $k - 1$, $2k - 1$, $4k - 5$, where k is a real number. Find the value of k and write down the first three terms.

21. The first three terms in an arithmetic sequence are $k + 6$, $2k + 1$, $k + 18$, where k is a real number. Find the value of k and write down the first four terms.

22. The first three terms in an arithmetic sequence are $k - 2$, $2k + 1$, $k + 14$, where k is a real number.

    (i) Find the value of k and write down the first four terms.

    (ii) Find, in terms of n, an expression for T_n and hence or otherwise, find T_{21}.

    (iii) Which term of the sequence is 243?

Patterns 3

If we are asked to investigate pictures containing a mixture of colours or shapes, the first step is easy – number the diagrams! From this we can establish a connection between the diagram number and amount of colours or shapes in each one. This should lead us to a numeric pattern or sequence.

PATTERNS, SEQUENCES AND SERIES

EXAMPLE

The diagram shows three shapes constructed using matchsticks.

  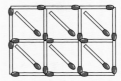

(i) If the pattern is continued, how many matchsticks will be needed for the 4th shape?
(ii) The amount of matchsticks used in the shapes form which type of sequence?
(iii) Find a formula, in terms of n, for the nth shape.
(iv) How many matchsticks will be needed for the 100th shape?

Solution:
Number the diagrams:

  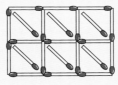
    1   2   3   4

Count the matchsticks in each diagram and put these in a table:

Shape	1	2	3	4
Number of matchsticks	9	16	23	?

(i) The difference between the number of matchsticks used is $16 - 9 = 7$ and $23 - 16 = 7$. As the differences are the same, we can easily predict the next number in the sequence. Therefore, the 4th shape will need $23 + 7 = 30$ matchsticks.

Alternatively, you could draw the new shape and simply count the matchsticks.

(ii) The numbers 9, 16, 23, . . . form an arithmetic sequence, as the terms have a common difference ($d = 7$).

(iii) $a = T_1 = 9, \quad d = 7.$

$$\begin{aligned} T_n &= a + (n-1)d \\ &= 9 + (n-1)(7) \\ &= 9 + 7n - 7 \\ &= 7n + 2 \end{aligned}$$

(iv) $$\begin{aligned} T_n &= 7n + 2 \\ T_{100} &= 7(100) + 2 \\ &= 702 \end{aligned}$$

Thus, 702 matchsticks will be needed for the 100th shape.

Exercise 5.6

1. These three diagrams were made using matches.

 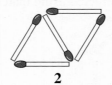 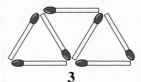

    1   2   3

  (i) If the pattern is continued, how many matches will be needed for the 4th, 5th and 6th diagrams?

  (ii) The amount of matches used in the diagrams form which type of sequence?

  (iii) Find a formula, in terms of n, for the nth diagram.

  (iv) How many matches will be needed for the 50th diagram?

  (v) Explain why there is no diagram with this pattern needing 200 matches.

2. The patterns shown are made from hexagonal tiles.

  Pattern 1   Pattern 2   Pattern 3

 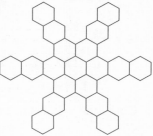

  (i) Construct a table showing the number of tiles for the first five patterns.

  (ii) Find a formula, in terms of n, for the number of tiles in the nth pattern.

  (iii) Which pattern is made from 253 tiles?

3. A series of shapes are made using matches.

  (i) How many matches will be needed for the 4th and 5th shapes?

  (ii) Find a formula, in terms of n, for the nth shape.

  (iii) Which shape will need 49 matches?

4. Orange and black discs are arranged in rectangular patterns as shown.

(i) How many discs will be in the 10th rectangle?

(ii) How many black discs will be in the 10th rectangle?

(iii) Describe two methods of finding the number of orange discs in the 10th rectangle.

(iv) Explain why there is no rectangle with this pattern needing 99 discs.

(v) Explain why there is no rectangle with this pattern needing 50 discs.

5. Rectangles containing red and white squares form a sequence as shown.

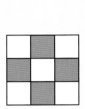 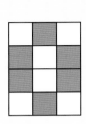 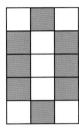

(i) Copy and complete the table.

Rectangle	1	2	3	4	5	6
Number of white squares						
Number of red squares						
Total number of squares						

(ii) Write down the number of white squares in the 10th rectangle.

(iii) Write down the formula, in terms of n, for the number of white squares in the nth rectangle.

(iv) Find a formula, in terms of n, for the number of red squares in the nth rectangle.

(v) Use this formula to find the number of red squares in the 10th rectangle.

(vi) Deduce the total number of squares in the 10th rectangle.

(vii) Find a formula, in terms of n, for the total number of squares by using your answers from (iii) and (iv).

(viii) Verify your formula from (vii) by letting $n = 2$ and checking your table from (i).

6. The set of diagrams have been made from straws.

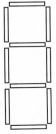 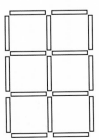 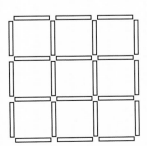

(i) How many straws will be needed for the 5th diagram?

(ii) Find a formula, in terms of *n*, for the *n*th diagram.

(iii) How many matches will be needed for the 20th diagram?

(iv) If $a : b$ is the ratio of the width to the height of the 1st diagram, find the value of *a* and the value of *b*.

(v) Which diagram has a ratio of its width to its height in the form $b : a$?

7. A long rectangular block is made using magnetic cubes. The sides are 1 cm. The diagram shows a block of four such cubes.

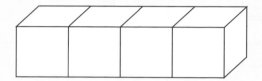

(i) How long is this block?

(ii) What is the surface area of this block?

(iii) If another block is attached to the end, what is the surface area?

(iv) What is the surface area of a block using 10 cubes?

(v) What is the surface area of a block using *n* cubes?

8. Patterns of dots are created as shown in this diagram.

(i) Draw the next two patterns.

(ii) Copy this table and complete it.

Diagram	1	2	3	4	5
Number of dots	5				

(iii) Find a formula, in terms of *n*, for the *n*th pattern.

(iv) Is it possible to predict from the table in (i) whether there is an even or an odd number of dots in the 50th diagram? Explain your answer.

(v) Is there an even or an odd number of dots in the 99th diagram?

(vi) Verify your predictions in (iv) and (v) by using the formula from (ii).

9. A shop stacks two of its products as shown. The Economy brand is shown as a yellow square while the Deluxe brand is purple.

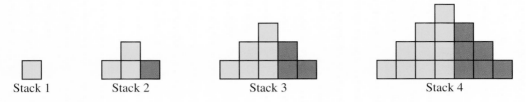

Stack 1   Stack 2   Stack 3   Stack 4

   (i) Write out, as a sequence, the number of Economy products in the first four stacks.

   (ii) Investigate whether the number of Economy products forms an arithmetic sequence.

   (iii) Write out, as a sequence, the number of Deluxe products in the first four stacks.

   (iv) Investigate whether the number of Deluxe products forms an arithmetic sequence.

10. A garden landscaper constructed small ponds of varying sizes. The diagram shows the three consecutive examples of a pond and the slabs that surround it.

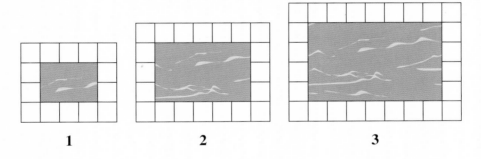

1   2   3

   (i) Write out, as a sequence, the number of slabs needed for these three ponds.

   (ii) Investigate whether the sequence forms an arithmetic sequence. Explain your conclusion.

   (iii) Find a formula, in terms of n, for the number of slabs needed to surround the nth pond.

   (iv) Which pond would require 98 slabs to surround it?

   (v) Write out, as a sequence, the width (shown vertically) as a number of slabs needed for these three ponds.

   (vi) Find a formula, in terms of n, for the width of the nth pond.

   (vii) What is the width and length of a pond which requires 98 slabs to surround it?

   (viii) Draw the small pond that would precede the three shown and count the number of slabs needed. Verify that it fits in with the sequence you found earlier.

Arithmetic sequence 2

In some questions we are given two terms of an arithmetic sequence. In this case, we use the method of simultaneous equations to find a and d.

EXAMPLE

In an arithmetic sequence, the fifth term, T_5, is 19 and the eighth term, T_8, is 31. Find the first term, a, and the common difference, d.

Solution:

We are given two equations in disguise and we use these to find a and d.

$$T_n = a + (n-1)d$$

Given: $T_5 = 19$
$\therefore a + 4d = 19$ ①

Given: $T_8 = 31$
$\therefore a + 7d = 31$ ②

Now solve the simultaneous equations ① and ② to find the value of a and the value of d.

$a + 7d = 31$ ②
$-a - 4d = -19$ ① × −1
─────────────
$3d = 12$
$d = 4$

Put $d = 4$ into ① or ②
$a + 4d = 19$ ①
$a + 4(4) = 19$
$a + 16 = 19$
$a = 3$

Having found d, the next step is to find a.

Thus, $a = 3$ and $d = 4$.

Exercise 5.7

1. In an arithmetic sequence, the third term, T_3, is 7 and the fifth term, T_5, is 11.
   Find the first term, a, and the common difference, d.
2. In an arithmetic sequence, the fifth term, T_5, is 13 and the eighth term, T_8, is 22.
   Find the first term, a, and the common difference, d.
3. In an arithmetic sequence, the fourth term, T_4, is 19 and the seventh term, T_7, is 31.
   Find the first term, a, and the common difference, d.

4. In an arithmetic sequence, the fifth term, T_5, is 23 and the ninth term, T_9, is 43.
   Find the first term, a, and the common difference, d.

5. In an arithmetic sequence, the sixth term, T_6, is 35 and the eighth term, T_8, is 47.
   Find the first term, a, and the common difference, d.

6. In an arithmetic sequence, the first term, T_1, is 7 and the fifth term, T_5, is 19.
   (i) Find the common difference, d.
   (ii) Find, in terms of n, an expression for T_n, the nth term, and hence or otherwise, find T_{20}.
   (iii) Which term of the sequence is 100?

7. In an arithmetic sequence, the sum of the third term, T_3, and the seventh term, T_7, is 38 and the sixth term, T_6, is 23.
   (i) Find the first term, a, and the common difference, d.
   (ii) Find, in terms of n, an expression for T_n, the nth term.
   (iii) Show that $T_{19} = 5T_4$.
   (iv) For what value of n is $T_n = 99$?

8. The first four terms of an arithmetic sequence are 5, p, q, 11.
   (i) Find the value of p and the value of q.
   (ii) Find T_{10}, the tenth term.

9. The first five terms of an arithmetic sequence are p, q, 4, r, -2.
   (i) Find the value of p, the value of q and the value of r.
   (ii) Find T_{20}, the twentieth term.

Arithmetic sequence 3

To verify that a sequence is arithmetic, we must show the following:

$$T_n - T_{n-1} = \text{constant}$$

To show that a sequence is **not arithmetic**, it is only necessary to show that the difference between any two consecutive terms is not the same.
In practice, this usually involves showing that $T_3 - T_2 \neq T_2 - T_1$ or similar.

EXAMPLE 1

T_n, the nth term, of a sequence is given by $T_n = 5n + 2$.
Verify that the sequence is arithmetic.

Solution:

$$T_n = 5n + 2$$
$$T_{n-1} = 5(n - 1) + 2 \quad \text{(replace } n \text{ with } (n - 1))$$
$$= 5n - 5 + 2$$
$$\therefore T_{n-1} = 5n - 3$$
$$T_n - T_{n-1} = 5n + 2 - (5n - 3)$$
$$= 5n + 2 - 5n + 3$$
$$T_n - T_{n-1} = 5$$
$$\therefore T_n - T_{n-1} = \text{a constant}$$

Thus, the sequence is arithmetic.

EXAMPLE 2

T_n, the nth term, of a sequence is given by $T_n = n^2 + 3n$.
Verify that the sequence is not arithmetic.

Solution:

$T_n = n^2 + 3n$

$T_1 = (1)^2 + 3(1)$
$= 1 + 3$
$T_1 = 4$

$T_2 = (2)^2 + 3(2)$
$= 4 + 6$
$T_2 = 10$

$T_3 = (3)^2 + 3(3)$
$= 9 + 9$
$T_3 = 18$

$$T_3 - T_2 = 18 - 10 = 8 \qquad T_2 - T_1 = 10 - 4 = 6$$
$$T_3 - T_2 \neq T_2 - T_1$$

Thus, the sequence is not arithmetic.

Note: We could also have shown $T_n - T_{n-1} \neq$ a constant to show that the sequence is not arithmetic.

Exercise 5.8
In questions 1–12, you are given the nth term, T_n, of a sequence. Show that the sequence is arithmetic.

1. $T_n = 2n + 3$
2. $T_n = 3n + 1$
3. $T_n = 4n + 5$
4. $T_n = 5n$
5. $T_n = 2n$
6. $T_n = 3n - 2$
7. $T_n = n - 4$
8. $T_n = 2n - 5$
9. $T_n = 7n + 1$
10. $T_n = 3 - n$
11. $T_n = 5 - 2n$
12. $T_n = 4 - 3n$

In questions 13–18, you are given the nth term, T_n, of a sequence. Show that the sequence is not arithmetic.

13. $T_n = n^2 + 2n$
14. $T_n = n^2 + 5n$
15. $T_n = n^2 - 3n$
16. $T_n = 2n^2 + n$
17. $T_n = 2n^2 - 1$
18. $T_n = n^2 - 4n + 3$

Series

When we add together the terms of a sequence, we get a series.

For example:

Sequence: 1, 4, 7, 10, ...

Series: $1 + 4 + 7 + 10 + \cdots$

The commas are replaced by plus signs to form the series.
The sum of the series is the result of adding the terms.
The sum of the first n terms of a series is denoted by S_n.

$$\therefore S_n = T_1 + T_2 + T_3 + \cdots + T_n$$

Note: Even though each term is separated by a plus sign rather than a comma, we still write $T_1 = 4$, $T_2 = 7$, $T_3 = 10$, etc.

From this we have:

$S_1 = T_1$
$S_2 = T_1 + T_2$
$S_3 = T_1 + T_2 + T_3$, etc.

Arithmetic series

An **arithmetic series** is the sum of the terms of an arithmetic sequence.

The sum of the first n terms of an arithmetic series is denoted by S_n.
The formula for S_n can be written in terms of n, a and d.

$$S_n = \frac{n}{2}[2a + (n - 1)d]$$

Note: Once we find a and d, we can answer any question about an arithmetic series.

EXAMPLE 1

The first three terms of an arithmetic series are $4 + 7 + 10 + \cdots$

(i) Find, in terms of n, an expression for S_n, the sum to n terms.

(ii) Find S_{20}, the sum of the first 20 terms.

Solution:

Note: Even though each term is separated by a plus sign rather than a comma, we still write $T_1 = 4$, $T_2 = 7$, $T_3 = 10$, etc.

(i) $4 + 7 + 10 + \cdots$

$$a = T_1 = 4 \qquad d = T_2 - T_1 = 7 - 4 = 3$$

$$S_n = \frac{n}{2}[2a + (n-1)d]$$

$$= \frac{n}{2}[2(4) + (n-1)(3)] \qquad \text{(put in } a = 4 \text{ and } d = 3\text{)}$$

$$= \frac{n}{2}[8 + 3n - 3]$$

$$S_n = \frac{n}{2}(3n + 5)$$

(ii) Find S_{20}.

Method 1:

$$S_n = \frac{n}{2}[2a + (n-1)d]$$

$$S_{20} = \frac{20}{2}[2(4) + (20-1)(3)]$$

(put in $n = 20$, $a = 4$ and $d = 3$)

$$S_{20} = 10(65)$$

$$S_{20} = 650$$

Method 2:

$$S_n = \frac{n}{2}(3n + 5) \quad \text{(from part (i))}$$

$$S_{20} = \frac{20}{2}(3(20) + 5)$$

$$= 10(65)$$

$$S_{20} = 650$$

In some questions we are given values of S_n and T_n for two values of n. In this case, we use the method of simultaneous equations to find a and d.

EXAMPLE 2

In an arithmetic series, the fifth term, T_5, is 14 and the sum of the first six terms, S_6, is 57.

(i) Find the first term a, and the common difference, d.

(ii) Show that: (a) $T_n = 3n - 1$ (b) $2S_n = 3n^2 + n$

Solution:

(i) We are given two equations in disguise and we use them to find a and d.

$T_n = a + (n-1)d$

Given: $T_5 = 14$

$\therefore a + (5-1)d = 14$

$a + 4d = 14$ ①

$S_n = \dfrac{n}{2}[2a + (n-1)d]$

Given: $S_6 = 57$

$\therefore \dfrac{6}{2}[2a + (6-1)d] = 57 \quad (n = 6)$

$3(2a + 5d) = 57$

$2a + 5d = 19$ ②

(divide both sides by 3)

Now solve the simultaneous equations ① and ② to find the value of a and the value of d.

$\begin{aligned} 2a + 8d &= 28 \quad &① \times 2 \\ -2a - 5d &= -19 \quad &② \times -1 \\ \hline 3d &= 9 \\ d &= 3 \end{aligned}$

Knowing $d = 3$, the next step is to find a.

Put $d = 3$ into ① or ②

$a + 4d = 14$ ①

$a + 4(3) = 14$

$a + 12 = 14$

$a = 2$

Thus, $a = 2$ and $d = 3$.

(ii) Show that $T_n = 3n - 1$ and $2S_n = 3n^2 + n$.

Replace a with 2 and d with 3 in the formulas for T_n and S_n.

(a) $T_n = a + (n-1)d$

$= 2 + (n-1)(3)$

(put in $a = 2$ and $d = 3$)

$T_n = 2 + 3n - 3$

$T_n = 3n - 1$

(b) $S_n = \dfrac{n}{2}[2a + (n-1)d]$

$= \dfrac{n}{2}[2(2) + (n-1)(3)]$

(put in $a = 2$ and $d = 3$)

$S_n = \dfrac{n}{2}[4 + 3n - 3]$

$S_n = \dfrac{n}{2}(3n + 1)$

$2S_n = n(3n + 1)$

(multiply both sides by 2)

$2S_n = 3n^2 + n$

In some questions we have to solve an equation in disguise to find the value of n.

> **EXAMPLE 3**
>
> Find the sum of the arithmetic series $7 + 9 + 11 + \cdots + 55$.
>
> **Solution:**
> The first step is to find the number of terms, n, that there are in the series.
>
> $a = T_1 = 1$ $\qquad\qquad\qquad\qquad\qquad\qquad d = T_2 - T_1 = 9 - 7 = 2$
>
> Equation given in disguise:
> Given: $\qquad T_n = 55$
> $\therefore a + (n - 1)d = 55$
> $\qquad 7 + (n - 1)(2) = 55$
> (put in $a = 7$ and $d = 2$)
> $\qquad 7 + 2n - 2 = 55$
> $\qquad\qquad 2n + 5 = 55$
> $\qquad\qquad\quad 2n = 50$
> $\qquad\qquad\quad\; n = 25$
>
> $S_n = \dfrac{n}{2}[2a + (n-1)d]$
>
> $S_{25} = \dfrac{25}{2}[2(7) + (25 - 1)(2)]$
>
> (put in $n = 25$, $a = 7$ and $d = 2$)
>
> $= \dfrac{25}{2}[14 + (24)(2)]$
>
> $= \dfrac{25}{2}[14 + 48]$
>
> $= \dfrac{25}{2}(62)$
>
> $S_{25} = 775$
>
> Thus, there are 25 terms.

Exercise 5.9

1. The first three terms of an arithmetic series are $3 + 5 + 7 + \cdots$.
   Find S_{10}, the sum of the first 10 terms.

2. The first three terms of an arithmetic series are $4 + 7 + 10 + \cdots$.
   Find S_{12}, the sum of the first 12 terms.

3. The first three terms of an arithmetic series are $1 + 5 + 9 + \cdots$.
   Find S_{18}, the sum of the first 18 terms.

4. The first three terms of an arithmetic series are $3 + 8 + 13 + \cdots$.
   (i) Find, in terms of n, an expression for S_n, the sum to n terms.
   (ii) Find S_{20}, the sum of the first 20 terms.

5. The first three terms of an arithmetic series are $10 + 13 + 16 + \cdots$.
   (i) Find, in terms of n, an expression for S_n, the sum to n terms.
   (ii) Find S_{30}, the sum of the first 30 terms.

6. The nth term of an arithmetic series is given by $2n + 3$.
   (i) Write down the first four terms.
   (ii) Write down the common difference.
   (iii) Find S_{16}, the sum of the first 16 terms.
7. In an arithmetic series, the fifth term, T_5, is 22 and the sum of the first four terms, S_4, is 38.
   (i) Find the first term, a, and the common difference, d.
   (ii) Show that: (a) $T_n = 5n - 3$ (b) $2S_n = 5n^2 - n$
8. In an arithmetic series, the eighth term, T_8, is 21 and the sum of the first six terms, S_6, is 18.
   (i) Find the first term, a, and the common difference, d.
   (ii) Show that: (a) $T_n = 4n - 11$ (b) $S_n = 2n^2 - 9n$
   (iii) Find: (a) T_{20} (b) S_{30}
9. In an arithmetic series, the seventh term, T_7, is 20 and the sum of the first five terms, S_5, is 40.
   (i) Find the first term, a, and the common difference, d.
   (ii) Show that: (a) $T_n = 3n - 1$ (b) $2S_n = 3n^2 + n$
   (iii) Find: (a) T_{30} (b) S_{30}
10. In an arithmetic series, the eighth term, T_8, is 27 and the sum of the first 10 terms, S_{10}, is 120.
    (i) Find the first term, a, and the common difference, d.
    (ii) Show that: (a) $T_n = 3(2n - 7)$ (b) $S_n = 3(n^2 - 6n)$
    (iii) Find: (a) T_{25} (b) S_6
11. In an arithmetic series, the sum of the first four terms, S_4, is 44 and the sum of the first six terms, S_6, is 102.
    (i) Find the first term, a, and the common difference, d.
    (ii) Find: (a) T_{20} (b) S_{20}
12. Find the sum of the arithmetic series $2 + 5 + 8 + \cdots + 59$.
13. Find the sum of the arithmetic series $1 + 5 + 9 + \cdots + 117$.
14. Find the sum of the arithmetic series $3 + 8 + 13 + \cdots + 248$.
15. The nth term of a series is given by $T_n = 3n + 2$.
    (i) Write down, in terms of n, an expression for T_{n-1}, the $(n - 1)$st term.
    (ii) Show that the series is arithmetic.
    (iii) Find S_{20}, the sum of the first 20 terms.

16. The first three terms of an arithmetic series are $10 + 20 + 30 + \cdots$.
   (i) Find, in terms of n, an expression for T_n, the nth term.
   (ii) Find, in terms of n, an expression for S_n, the sum to n terms.
   (iii) Using your expression for S_n, find the sum of the natural numbers that are both multiples of 10 and smaller than 2,001.

17. The first three terms in an arithmetic series are $1 + 3 + 5 + \cdots$.
   (i) Show that: (a) $T_n = 2n - 1$ (b) $S_n = n^2$
   (ii) Hence or otherwise, evaluate: (a) T_{20} (b) S_{20}
   (iii) How many terms need to be added to give a sum of 225?

18. The first three terms of an arithmetic series are $3a + 4a + 5a + \cdots$ where a is a real number.
   (i) Find, in terms of a, an expression for T_{10}, the tenth term.
   (ii) Find, in terms of a, an expression for S_{10}, the sum of the first 10 terms.
   (iii) If $S_{10} - T_{10} = 126$, find the value of a.
   (iv) Write down the first four terms of the series.
   (v) Write down, in terms of n, expressions for: (a) T_n (b) S_n
   (vi) Hence or otherwise, evaluate: (a) T_{20} (b) S_{20}

19. The general term, T_n, of an arithmetic series is given by $T_n = 2n + 5$.
   (i) Find the first term, a, and the common difference, d.
   (ii) For what value of n is the sum of the first n terms, S_n, equal to 160?

20. A pupil saves money each week. The pupil saves 40c in the first week, 60c in the second week, 80c in the next week, continuing this pattern for 50 weeks.
   (i) How much will the pupil save in the 50th week?
   (ii) How much will the pupil have saved after 50 weeks?

21. A woman accepted a post with a starting salary of €30,000. In each following year she received an increase in salary of €2,000. What were her total earnings in the first 12 years?

22. In a potato race, 10 potatoes are placed 8 m apart in a straight line. The object of the race is to pick up the first potato and place it in a basket 20 m in front of the first potato, then run to the second potato, pick it up and place it in the basket and so on.

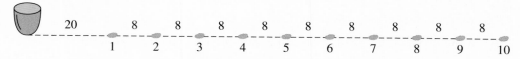

The race begins at the basket. Find the total distance covered by a contestant who finishes the race.

23. A display in a grocery store will consist of cans stacked as shown.

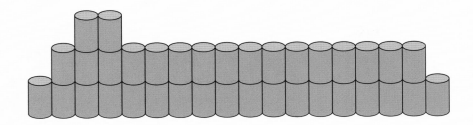

The first row is to have 18 cans and each row after the first is to have two cans fewer than the preceding row. How many cans will be needed in the display?

24. A man is given an interest-free loan. He repays the loan in monthly instalments. He repays €40 at the end of the first month, €44 at the end of the second month, €48 at the end of the third month, continuing the pattern of increasing the monthly repayments by €4 a month until the loan is repaid. The final monthly repayment is €228.

   (i) Show that it will take the man 48 months to repay the loan.

   (ii) Calculate the amount of the loan.

25. A ball rolls down a slope. The distances it travels in successive seconds are 2 cm, 6 cm, 10 cm, 14 cm, and so on. How many seconds elapse before it has travelled a total of 18 m?

Given S_n of an arithmetic series in terms of n

In many problems we are given an expression for S_n in terms of n and we need to find a and d. In this type of problem, we use the fact that for all arithmetic series:

$$T_n = S_n - S_{n-1} \quad \text{and} \quad T_1 = S_1$$

Examples: $T_2 = S_2 - S_1, \quad T_3 = S_3 - S_2, \quad T_9 = S_9 - S_8,$ and so on.

NEW CONCISE PROJECT MATHS 3 STRANDS 3, 4 & 5

 EXAMPLE 1

The sum of the first n terms, S_n, of an arithmetic series is given by $S_n = 2n^2 + n$. Find the first term, a, and the common difference, d.

Solution:

$$S_n = 2n^2 + n$$

$$S_1 = 2(1)^2 + (1)$$
$$= 2(1) + 1$$
$$= 2 + 1$$
$$= 3$$
$$a = T_1 = S_1 = 3$$

$$S_2 = 2(2)^2 + (2)$$
$$= 2(4) + 2$$
$$= 10$$
$$T_2 = S_2 - S_1$$
$$= 10 - 3$$
$$= 7$$
$$d = T_2 - T_1 = 7 - 3 = 4$$

Thus, $a = 3$ and $d = 4$.

 EXAMPLE 2

The general term, T_n, of an arithmetic sequence is given by $T_n = 2n + 5$. Find the first term, a, and the common difference, d. For what value of n is $S_n = 160$?

Solution:

$$T_n = 2n + 5$$
$$T_1 = 2(1) + 5 = 2 + 5 = 7 = a$$
$$T_2 = 2(2) + 5 = 4 + 5 = 9$$

$$d = T_2 - T_1$$
$$= 9 - 7$$
$$d = 2$$

Thus, $a = 7$ and $d = 2$.
Equation given in disguise:

$$S_n = 160$$
$$\frac{n}{2}[2a + (n-1)d] = 160 \quad \text{(we know } a \text{ and } d\text{, find } n\text{)}$$
$$\frac{n}{2}[2(7) + (n-1)(2)] = 160 \quad \text{(put in } a = 7 \text{ and } d = 2\text{)}$$
$$\frac{n}{2}(14 + 2n - 2) = 160$$
$$\frac{n}{2}(2n + 12) = 160$$
$$n(2n + 12) = 320 \quad \text{(multiply both sides by 2)}$$
$$2n^2 + 12n = 320$$
$$2n^2 + 12n - 320 = 0$$
$$n^2 + 6n - 160 = 0 \quad \text{(divide both sides by 2)}$$

$$(n - 10)(n + 16) = 0$$
$$n - 10 = 0 \text{ or } n + 16 = 0$$
$$n = 10 \text{ or } n = -16 \quad (\text{reject } n = -16)$$

Thus, $n = 10$.

Note: If n is a fraction or a negative number, reject it.

Exercise 5.10

1. The sum of the first n terms, S_n, of an arithmetic series is given by $S_n = n^2 + 2n$. Use S_1 and S_2 to find the first term, a, and the common difference, d.

2. The sum of the first n terms, S_n, of an arithmetic series is given by $S_n = n^2 + 3n$. Find the first term, a, and the common difference, d.

3. The sum of the first n terms, S_n, of an arithmetic series is given by $S_n = 3n^2 - 2n$. Find the first term, a, and the common difference, d.

4. The sum of the first n terms, S_n, of an arithmetic series is given by $S_n = 2n^2 - 3n$. Find the first term, a, and the common difference, d.

5. The sum of the first n terms, S_n, of an arithmetic series is given by $S_n = \dfrac{n(3n + 1)}{2}$.
   (i) Calculate the first term of the series.
   (ii) By calculating S_8 and S_7, find T_8, the eighth term of the series.

6. The sum of the first n terms, S_n, of an arithmetic series is given by $S_n = 2n^2 + n$.
   (i) Calculate the first term of the series and the common difference.
   (ii) Find, in terms of n, an expression for T_n, the nth term.
   (iii) Hence, calculate T_{10}.
   (iv) Show that $T_{10} = S_{10} - S_9$.

7. The sum of the first n terms, S_n, of an arithmetic series is given by $S_n = 2n^2 - 4n$.
   (i) Find the first term, a, and the common difference, d.
   (ii) Find, in terms of n, an expression for T_n, the nth term.
   (iii) Find T_{20} and verify that $T_{20} = S_{20} - S_{19}$.
   (iv) Starting with the first term, how many terms of the series must be added to give a sum of 160?

Differences – 2

Earlier, we investigated differences between differences on some sequences. While these are not arithmetic sequences, it is still possible to find an expression for T_n, the nth term. When it takes *two* differences to see a constant, the formula for T_n will be of the form $T_n = an^2 + bn + c$.

Furthermore, it can be shown that when the second differences are 2, then $T_n = n^2 + bn + c$, and when the second differences are 4, then $T_n = 2n^2 + bn + c$.

Here is the sequence 3, 6, 11, 18, 27, . . .

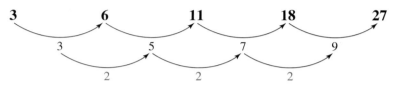

The general term of this sequence will be of the form $T_n = n^2 + bn + c$.

EXAMPLE

Find the general term, T_n, of the sequence 1, 5, 11, 19, 29, $\cdots$

Solution:

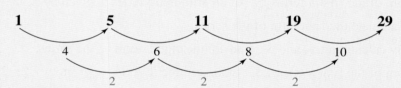

Because the second difference is 2, $T_n = n^2 + bn + c$. Now we need to find the value of b and the value of c.

We will use the first two terms to form two equations involving b and c and then use the method of simultaneous equations.

Given: $T_1 = 1$ and $T_2 = 5$

$T_n = n^2 + bn + c$
$T_1 = 1^2 + b(1) + c = 1$
$1 + b + c = 1$
$b + c = 0$ ①

$T_2 = 2^2 + b(2) + c = 5$
$4 + 2b + c = 5$
$2b + c = 1$ ②

$-b - c = 0 \quad\quad ① \times -1$
$\underline{2b + c = 1} \quad\quad ②$
$b = 1$

$b + c = 0$
$1 + c = 0$
$c = -1$

Thus, $T_n = n^2 + 1n - 1 = n^2 + n - 1$.

Exercise 5.11

In questions 1–6, find an expression for T_n for the following sequences.

1. 3, 7, 13, 21, 31, ...
2. 4, 9, 16, 25, 36, ...
3. 2, 6, 12, 20, 30, ...
4. 2, 5, 10, 17, 26, ...
5. 0, 0, 2, 6, 12, ...
6. 8, 15, 24, 35, 48, ...

7. (i) Find T_n of the sequence 0, 4, 10, 18, 28, ...

   (ii) Find T_{10}.

   (iii) Investigate whether T_{100} is less than or greater than 1,000.

8. A rocket is launched from ground level and its height is recorded each second after blast-off. The heights are 22 m, 36 m, 52 m, 70 m, ...

   (i) Find T_n of the sequence.

   (ii) How high will the rocket be after 10 seconds?

   (iii) How high will the rocket be after 1 minute?

   (iv) Noting that the rocket does not rise uniformly, give a reason why a rocket would rise faster as time passes.

9. Mary is making rectangular patterns from counters. The length of each rectangle is one more than the width. The first three patterns are shown.

   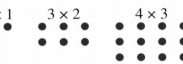

   (i) Each time, to get the next pattern, she increases the width by one more counter. Write an expression for the number of counters in the nth pattern.

   (ii) Mary notices that there is also a pattern in how many extra counters she needs each time to make the next pattern. How many extra counters does she need to make the $(n + 1)$th pattern from the nth pattern? Write your answer in the form $a(n + b)$.

CHAPTER 6 FUNCTIONS

Terminology and Notation

A function is a rule that changes one number (input) into another number (output). Functions are often represented by the letters f, g, h or k. We can think of a function, f, as a number machine which changes an input, x, into an output, $f(x)$.

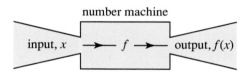

$f(x)$, which denotes the output, is read as 'f of x'.

For example, let's represent the function 'double input and then add 5' by the letter f.

This can be written as:

$$f : x \rightarrow 2x + 5 \quad \text{or} \quad f(x) = 2x + 5 \quad \text{or} \quad y = 2x + 5$$

$$(\text{input, output}) = (x, f(x)) = (x, 2x + 5) = (x, y)$$

Note: A **function** is also called a '**mapping**' or simply a '**map**'.

One number is mapped onto another number.
In the above example, x is mapped onto $2x + 5$, usually written $f : x \rightarrow 2x + 5$.

Input number

If $f : x \rightarrow 2x + 5$, then $f(3)$ means 'input 3 into the function',
i.e. it is the result of applying the function f to the number 3.

$$f(3) = 2(3) + 5 = 6 + 5 = 11 \quad (\text{input} = 3, \text{output} = 11)$$

$$(\text{input, output}) = (3, f(3)) = (3, 11)$$

> A function does exactly the same to each input number, and produces only one output number for each input number.

The set of numbers that are put into a function is called the '**domain**'.
The set of numbers that comes out of a function is called the '**range**'.
A function connects **every** input in the domain to an input in the range.
A function is another way of writing an algebraic formula that links input to output.

FUNCTIONS

Example

The function g is defined as $g : x \rightarrow 3x - 2$, $\quad x \in \mathbf{R}$.

(i) Find: (a) $g(6)$ (b) $g(\frac{4}{3})$.

(ii) Find the number k such that $kg(\frac{4}{3}) = g(6)$.

(iii) Find the value of x for which $g(x) = 13$.

Solution:

$$g(x) = 3x - 2$$

(i) (a) $g(6) = 3(6) - 2$
$ = 18 - 2$
$ = 16$

(b) $g(\frac{4}{3}) = 3(\frac{4}{3}) - 2$
$\phantom{g(\frac{4}{3})} = 4 - 2$
$\phantom{g(\frac{4}{3})} = 2$

(ii) $kg(\frac{4}{3}) = g(6)$
$k(2) = 16$
$2k = 16$
$k = 8$

(iii) $g(x) = 13$
$3x - 2 = 13$
$3x = 15$
$x = 5$

Example

$f : x \rightarrow x^2 - x$ and $g(x) = x + 1$.

(i) Evaluate: (a) $f(-1)$ (b) $g(-\frac{1}{4})$.

(ii) Find the two values of x for which $2f(x) = 3g(x)$.

Solution:

(i) (a) $f(x) = x^2 - x$
$f(-1) = (-1)^2 - (-1)$
$ = 1 + 1$
$ = 2$

(b) $g(x) = x + 1$
$g(-\frac{1}{4}) = -\frac{1}{4} + 1$
$\phantom{g(-\frac{1}{4})} = \frac{3}{4}$

(ii) $2f(x) = 3g(x)$
$2(x^2 - x) = 3(x + 1)$
$2x^2 - 2x = 3x + 3$
$2x^2 - 2x - 3x - 3 = 0$
$2x^2 - 5x - 3 = 0$
$(2x + 1)(x - 3) = 0$
$2x + 1 = 0 \quad \text{or} \quad x - 3 = 0$
$2x = -1 \quad \text{or} \quad x = 3$
$x = -\frac{1}{2} \quad \text{or} \quad x = 3$

Example

Let $f(x) = \dfrac{1}{x+3}$, $x \in \mathbf{R}$.

Find: (i) $f(-\tfrac{10}{3})$ (ii) $f(-\tfrac{1}{2})$ (iii) $f(\tfrac{4}{5})$.

For what real value of $f(x)$ is $f(x)$ not defined?

Solution:

$$f(x) = \dfrac{1}{x-3}$$

(i) $f(-\tfrac{10}{3}) = \dfrac{1}{-\tfrac{10}{3}+3}$

$= \dfrac{3}{-10+9}$

(multiply top and bottom by 3)

$= \dfrac{3}{-1}$

$= -3$

(ii) $f(-\tfrac{1}{2}) = \dfrac{1}{-\tfrac{1}{2}+3}$

$= \dfrac{2}{-1+6}$

(multiply top and bottom by 2)

$= \dfrac{2}{5}$

(iii) $f(\tfrac{4}{5}) = \dfrac{1}{\tfrac{4}{5}+3}$

$= \dfrac{5}{4+15}$

(multiply top and bottom by 5)

$= \dfrac{5}{19}$

Division by 0 (zero) is not defined for real numbers.

When $x = -3$, $\dfrac{1}{x+3} = \dfrac{1}{-3+3} = \dfrac{1}{0}$ which is undefined.

∴ $x = -3$ is the real value for which $f(x)$ is undefined.

Exercise 6.1

1. The function f is defined as $f : x \to 3x + 4$, $x \in \mathbf{R}$.
   Find: (i) $f(2)$ (ii) $f(5)$ (iii) $f(-1)$ (iv) $f(0)$ (v) $f(\tfrac{2}{3})$

2. The function g is defined as $g : x \to 5x - 2$, $x \in \mathbf{R}$.
   Find: (i) $g(3)$ (ii) $g(-2)$ (iii) $g(0)$ (iv) $g(-1)$ (v) $g(\tfrac{2}{5})$
   Find the value of x for which $g(x) = 18$.

3. The function f is defined as $f : x \to x^2 + 3x$, $x \in \mathbf{R}$.
   Find: (i) $f(3)$ (ii) $f(1)$ (iii) $f(0)$ (iv) $f(-2)$ (v) $3f(-4)$
   Find the two values of x for which $f(x) = 10$.

4. The function f is defined as $f : x \to 4 - 3x$, $x \in \mathbf{R}$.
   (i) Find: (a) $f(-2)$ (b) $f(-\tfrac{1}{3})$
   (ii) Find the number k such that $f(-2) = kf(-\tfrac{1}{3})$.

5. The function f is defined as $f : x \to 7 - 4x$, $\quad x \in \mathbf{R}$.
   Find the number k such that $kf(-\frac{3}{2}) = f(-8)$.

6. The function f is defined as $f : x \to x(x + 2)$, $\quad x \in \mathbf{R}$.
   Find: (i) $f(1)$ (ii) $f(-1)$ (iii) $f(2) + f(-2)$ (iv) $-2f(-3)$.
   Find the two values of x for which $f(x) = 15$.

7. The function f is defined as $f : x \to x^2 + 2x - 1$, $\quad x \in \mathbf{R}$.
   Find the value of: (i) $f(0)$ (ii) $f(-1)$ (iii) $f(\frac{1}{2})$ (iv) $f(\frac{3}{5})$.
   Find the two values of k for which $f(k) = -1$.

8. The function f is defined as $f : x \to \dfrac{1}{x - 1}$, $\quad x \in \mathbf{R}, x \neq 1$.
   Evaluate: (i) $f(3)$ (ii) $f(0)$ (iii) $f(\frac{3}{2})$.
   For what real value of x is $f(x)$ not defined?

9. The function f is defined as $f : x \to \dfrac{1}{x + 4}$, $\quad x \in \mathbf{R}$.
   Evaluate: (i) $f(1)$ (ii) $f(-2)$ (iii) $f(-\frac{4}{5})$, writing your answers as decimals.
   For what real value of x is $f(x)$ not defined?
   Solve for x: $\quad f(x) = \frac{1}{3}$.

10. The function f is defined as $f : x \to x^2 - 1$, $\quad x \in \mathbf{R}$.
    (i) Find: (a) $f(4)$ (b) $f(-3)$ (c) $f(2)$.
    (ii) For what value of $k \in \mathbf{R}$ is $2k + 1 + f(4) = f(-3)$?
    (iii) Find the values of x for which $f(x) - f(2) = 0$.
    (iv) Verify that $f(x - 1) = f(1 - x)$.

11. $f : x \to x^2 + 1$ and $g : x \to 2x$ are two functions defined on $\mathbf{R}$.
    (i) Find $f(\sqrt{3})$ and $g(1)$.
    (ii) Find the value of k for which $f(\sqrt{3}) = kg(1)$.
    (iii) Find the value of x for which $f(x) = g(x)$.
    (iv) Verify that $f(x + 2) = g(x^2 + 2x + 1) - f(x) + f(\sqrt{3})$.

12. $f : x \to \sqrt{x}$ and $g : x \to \dfrac{x}{1 - x}$ are two functions defined on $\mathbf{R}$.
    Verify that $f(\frac{1}{4}) > g(\frac{1}{4})$.

Functions with Missing Coefficients

In some questions coefficients of the functions are missing and we are asked to find them. In this type of question we are given equations in disguise, and by solving these equations we can calculate the missing coefficients.

Notation

$f(x) = y$
$f(2) = 3$ means when $x = 2$, $y = 3$, or the point $(2, 3)$ is on the graph of the function.
$f(-1) = 0$ means when $x = -1$, $y = 0$, or the point $(-1, 0)$ is on the graph of the function.

Example

$f : x \to 3x + k$ and $g : x \to x^2 + hx - 8$ are two functions defined on **R**.

(i) If $f(-2) = 1$, find the value of k.

(ii) If $g(-3) = -5$, find the value of h.

Solution:

(i) $f(x) = 3x + k$

Given: $f(-2) = 1$
$$3(-2) + k = 1$$
$$-6 + k = 1$$
$$k = 7$$

(ii) $g(x) = x^2 + hx - 8$

Given: $g(-3) = -5$
$$\therefore (-3)^2 + h(-3) - 8 = -5$$
$$9 - 3h - 8 = -5$$
$$-3h + 1 = -5$$
$$-3h = -6$$
$$3h = 6$$
$$h = 2$$

Example

$g : x \to ax^2 + bx - 3$ is a function defined on **R**.
If $g(2) = 15$ and $g(-1) = -6$, write down two equations in a and b.
Hence, calculate the value of a and the value of b.

Solution:

$$g(x) = ax^2 + bx - 3$$

Given: $g(2) = 15$
$$\therefore a(2)^2 + b(2) - 3 = 15$$
$$a(4) + b(2) - 3 = 15$$
$$4a + 2b - 3 = 15$$
$$4a + 2b = 18$$
$$2a + b = 9 \quad ①$$

Given: $g(-1) = -6$
$$\therefore a(-1)^2 + b(-1) - 3 = -6$$
$$a(1) + b(-1) - 3 = -6$$
$$a - b - 3 = -6$$
$$a - b = -3 \quad ②$$

We now solve between equations ① and ②

$$2a + b = 9 \quad ①$$
$$\underline{a - b = -3 \quad ②}$$
$$3a = 6 \quad \text{(add)}$$
$$a = 2$$

put $a = 2$ into ① or ②

$$2a + b = 9 \quad ①$$
$$\downarrow$$
$$2(2) + b = 9$$
$$4 + b = 9$$
$$b = 5$$

Thus, $a = 2$ and $b = 5$.

FUNCTIONS

Example ▼

The graph of the quadratic function
$f: x \to x^2 + px + q, \quad x \in \mathbf{R}$, is shown.
Find the value of p and the value of q.

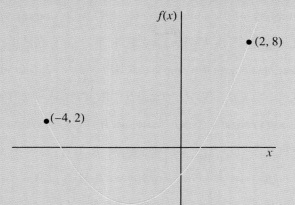

Solution:

$$f(x) = x^2 + px + q$$

The graph goes through the point $(-4, 2)$

∴ **Given:** $\qquad f(-4) = 2$
∴ $\qquad (-4)^2 + p(-4) + q = 2$
$\qquad 16 - 4p + q = 2$
$\qquad -4p + q = -14$
$\qquad 4p - q = 14 \qquad$ ①

The graph goes through the point $(2, 8)$

∴ **Given:** $\qquad f(2) = 8$
∴ $\qquad (2)^2 + p(2) + q = 8$
$\qquad 4 + 2p + q = 8$
$\qquad 2p + q = 4 \qquad$ ②

We now solve between equations ① and ②:

$\qquad 4p - q = 14 \qquad$ ①
$\qquad \underline{2p + q = 4} \qquad$ ②
$\qquad 6p = 18 \qquad$ (add)
$\qquad p = 3$

put $p = 3$ into ① or ②

$\qquad 2p + q = 4 \qquad$ ②
$\qquad \downarrow$
$\qquad 2(3) + q = 4$
$\qquad 6 + q = 4$
$\qquad q = -2$

Thus, $p = 3 \quad$ and $\quad q = -2$.

Alternatively, let $y = f(x)$, i.e., $y = x^2 + px + q$.

$(-4, 2)$ is on the graph of the curve

∴ $\quad 2 = (-4)^2 + p(-4) + q$
(put in $x = -4$, $y = 2$)
$\qquad 2 = 16 - 4p + q$
$\qquad -14 = -4p + q$
$\qquad 14 = 4p - q \qquad$ ①

$(2, 8)$ is on the graph of the curve

∴ $\quad 8 = (2)^2 + p(2) + 1$
(put in $x = 2$, $y = 8$)
$\qquad 8 = 4 + 2p + q$
$\qquad 4 = 2p + q \qquad$ ②

Then solve the simultaneous equations ① and ② as before, to get $p = 3$ and $q = -2$.

Exercise 6.2

1. Let $f(x) = 5x + k$ where $k \in \mathbf{R}$. If $f(1) = 7$, find the value of k.

2. Let $g(x) = 3x + h$ where $h \in \mathbf{R}$. If $g(-1) = 1$, find the value of h.

3. Let $h(x) = ax + 7$ where $a \in \mathbf{R}$. If $h(-3) = -2$, find the value of a.

4. Let $k(x) = x^2 - 3x + b$ where $b \in \mathbf{R}$. If $k(-1) = -6$, find the value of b.

5. Let $f(x) = ax^2 + 3x$ where $a \in \mathbf{R}$. If $f(-1) = -1$, find the value of a.

6. Let $f(x) = (x + k)(x + 3)$ where $k \in \mathbf{R}$. If $f(-2) = -6$, find the value of k.

7. Let $g(x) = a^2x^2 - 7ax + 6$ where $a \in \mathbf{R}$. If $g(2) = -6$, find two possible values of a.

8. Let $f(x) = x^2 + ax + b$ where $a, b \in \mathbf{R}$.
   (i) Find the value of a, given that $f(2) = f(-4)$.
   (ii) If $f(-5) = 12$, find the value of b.

9. $f : x \rightarrow 2x + a$ and $g : x \rightarrow 3x + b$.
   If $f(2) = 7$ and $g(1) = -1$, find the value of a and the value of b.

10. $h : x \rightarrow 2x + a$ and $k : x \rightarrow b - 5x$ are two functions defined on $\mathbf{R}$.
    If $h(1) = -5$ and $k(-1) = 4$, find the value of a and the value of b.

11. $f : x \rightarrow 3x + a$ and $g : x \rightarrow ax + b$ are two functions defined on $\mathbf{R}$.
    If $f(2) = 8$ and $g(2) = 1$:
    (i) find the value of a and the value of b
    (ii) find $f(-1)$ and $g(4)$
    (iii) using your values of a and b from (i), find the two values of x for which:
    $ax^2 - (a - b)x + 2ab = 0$.

12. $h : x \rightarrow 2x - a$ and $k : x \rightarrow ax + b$ are two functions defined on $\mathbf{R}$, where a and $b \in \mathbf{Z}$.
    $h(3) = 1$ and $k(5) = 8$.
    (i) Find the value of a and the value of b.
    (ii) Hence, list the values of x for which $h(x) \geq k(x)$, $x \in \mathbf{N}$.

13. $g : x \rightarrow ax^2 + bx + 1$ is a function defined on $\mathbf{R}$.
    If $g(1) = 2$ and $g(-1) = 6$, write down two equations in a and b.
    Hence, calculate the value of a and the value of b.

14. $g : x \rightarrow px^2 + qx - 3$ is a function defined on $\mathbf{R}$.
    If $g(1) = 4$ and $g(-1) = -6$, write down two equations in p and q.
    Hence, calculate the value of p and the value of q.
    Find the two values of x for which $px^2 + qx - 3 = 0$.

15. $g : x \rightarrow ax^2 + bx + 1$ is a function defined on $\mathbf{R}$.
    (i) If $g(1) = 0$ and $g(2) = 3$, write down two equations in a and b.
    (ii) Hence, calculate the value of a and the value of b.
    (iii) Using your values of a and b from (ii), find the two values of x for which $ax^2 + bx = bx^2 + ax$.

16. $f: x \rightarrow ax^2 + bx + c$, where a, b and c are real numbers.
    If $f(0) = -3$, find the value of c.
    If $f(-1) = 6$ and $f(2) = 3$, find the value of a and the value of b.

17. $h: x \rightarrow x^2 + x + q$ is a function defined on **R**, where $q \in \mathbf{Z}$.
    (i) If $h(-3) = 0$, find the value of q.
    (ii) Hence, solve the equation $h(x + 5) = 0$.

18. The graph of the linear function
    $f: x \rightarrow ax + b$, $x \in \mathbf{R}$ is shown.
    Find the values of a and b.

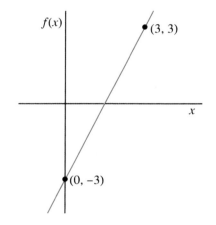

19. The graph of the quadratic function
    $f: x \rightarrow x^2 + bx + c$, $x \in \mathbf{R}$, is shown.
    Find the values of b and c.
    Hence, find the value of k.

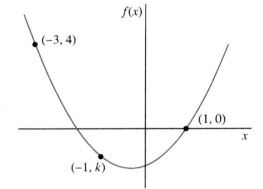

20. The graph of the quadratic function
    $g: x \rightarrow x^2 + px + q$, $x \in \mathbf{R}$, is shown.
    Find the values of p and q.
    Hence, find the values of a, b and c.
    Solve the equation:

    $$\frac{c-1}{b} = \frac{1}{x} + \frac{1}{x+p}$$

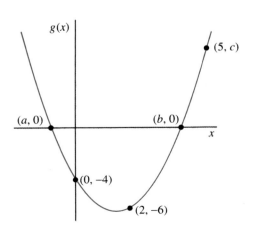

21. The graph of the quadratic function
$g : x \to ax^2 + bx - 3$, $x \in \mathbf{R}$, is shown.
Find the values of a and b.
Hence, calculate the value of h and k.
Solve the equation:

$$\frac{1}{x} + \frac{1}{x+k} + h = 0$$

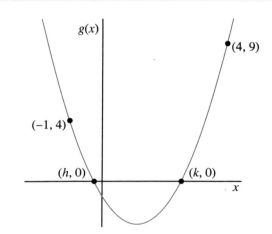

22. The graph of the quadratic function
$h : x \to c + bx - x^2$, $x \in \mathbf{R}$, is shown.
Find the value of b and c.
$k : x \to px + q$ is a function defined on $\mathbf{R}$.
If $k(0) = -1$ and $k(1) = 1$, find the value of p and the value of q.
Hence, find the two values of x for which $k(x) = h(x)$.

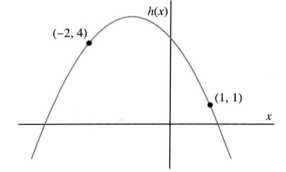

23. Let $f(x) = x^2 + bx + c$, $x \in \mathbf{R}$.
The solutions of $f(x) = 0$ are -5 and 2.
Find the value of b and the value of c.
If $f(-3) = k$, find the value of k.
Solve the equation $f(x) - k = 0$.

24. Let $g(x) = x^2 + bx + c$, $x \in \mathbf{R}$.
The solutions of $g(x) = 0$ are symmetrical about the line $x = -1$.
If $x = 1$ is one solution of $g(x) = 0$, find the other solution.
Find the value of b and the value of c.

CHAPTER 7: GRAPHING FUNCTIONS

Notation

The notation $y = f(x)$ means 'the value of the output y depends on the value of the input x, according to some rule called f'. Hence, y and $f(x)$ are interchangeable, and the y-axis can also be called the $f(x)$-axis.

Note: It is very important not to draw a graph outside the given values of x.

Graphing Linear Functions

The first four letters in the word '**linear**' spell '**line**'. Therefore the graph of a linear function will be a straight line. A linear function is usually given in the form $f : x \rightarrow ax + b$, where $a \neq 0$ and a, b are constants. For example, $f : x \rightarrow 2x + 5$. As the graph is a straight line, two points are all that is needed to graph it. In the question, you will always be given a set of inputs, x, called the **domain**.

To graph a linear function do the following:

> 1. Choose two suitable values of x, in the given domain.
>   (Two suitable values are the smallest and largest values of x.)
> 2. Substitute these in the function to find the two corresponding values of y.
> 3. Plot the points and draw the line through them.

Note: $-3 \leq x \leq 2$ means 'x is between -3 and 2, including -3 and 2'.

Example ▼

Graph the function $f : x \rightarrow 2x + 1$, in the domain $-3 \leq x \leq 2$, $x \in \mathbf{R}$.

Let $y = f(x) \Rightarrow y = 2x + 1$.

x	$2x + 1$	y
-3	$-6 + 1$	-5
2	$4 + 5$	5

Plot the points $(-3, -5)$ and $(2, 5)$ and join them with a straight line.

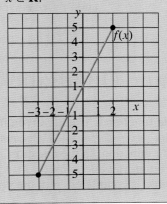

Exercise 7.1

Graph each of the following functions in the given domain:

1. $f : x \rightarrow 2x + 3$ in the domain $-3 \leqslant x \leqslant 2$, $x \in \mathbf{R}$
2. $g : x \rightarrow 3x + 2$ in the domain $-2 \leqslant x \leqslant 3$, $x \in \mathbf{R}$
3. $f : x \rightarrow 4x + 3$ in the domain $-3 \leqslant x \leqslant 3$, $x \in \mathbf{R}$
4. $g : x \rightarrow 2x - 5$ in the domain $-1 \leqslant x \leqslant 5$, $x \in \mathbf{R}$
5. $h : x \rightarrow x - 2$ in the domain $-2 \leqslant x \leqslant 5$, $x \in \mathbf{R}$
6. $f : x \rightarrow 3x - 1$ in the domain $-3 \leqslant x \leqslant 3$, $x \in \mathbf{R}$
7. $g : x \rightarrow 5x - 2$ in the domain $-1 \leqslant x \leqslant 2$, $x \in \mathbf{R}$
8. $k : x \rightarrow x$ in the domain $-3 \leqslant x \leqslant 3$, $x \in \mathbf{R}$
9. $f : x \rightarrow 2x$ in the domain $-2 \leqslant x \leqslant 2$, $x \in \mathbf{R}$
10. $g : x \rightarrow 3x$ in the domain $-3 \leqslant x \leqslant 2$, $x \in \mathbf{R}$
11. $f : x \rightarrow -x$ in the domain $-3 \leqslant x \leqslant 3$, $x \in \mathbf{R}$
12. $h : x \rightarrow 2 - x$ in the domain $-4 \leqslant x \leqslant 4$, $x \in \mathbf{R}$
13. $k : x \rightarrow 3 - 2x$ in the domain $-3 \leqslant x \leqslant 4$, $x \in \mathbf{R}$
14. $f : x \rightarrow 4 - 3x$ in the domain $-2 \leqslant x \leqslant 4$, $x \in \mathbf{R}$
15. $g : x \rightarrow -1 - x$ in the domain $-4 \leqslant x \leqslant 3$, $x \in \mathbf{R}$

16. Using the same axis and scales, graph the functions:

    $f : x \rightarrow x - 6$, $g : x \rightarrow -2x$, in the domain $-1 \leqslant x \leqslant 5$, $x \in \mathbf{R}$.

    (i) From your graph, write down the coordinates of the point of intersection of f and g.
    (ii) Verify your answer to part (i) by solving the simultaneous equations:
    $x - y = 6$ and $2x + y = 0$.

Graphing Quadratic Functions

A **quadratic** function is usually given in the form $f : x \rightarrow ax^2 + bx + c$, $a \neq 0$, and a, b, c are constants. For example, $f : x \rightarrow 2x^2 - x + 3$. Because of its shape, quite a few points are needed to plot the graph of a quadratic function. In the question, you will always be given a set of inputs, x, called the **domain**. With these inputs, a table is used to find the corresponding set of outputs, y or $f(x)$, called the **range**. When the table is completed, plot the points and join them with a '**smooth curve**'.

GRAPHING FUNCTIONS

Notes on making out the table:

1. Work out each column separately, i.e. all the x^2 values first, then all the x values, and finally the constant. (Watch for patterns in the numbers.)

2. Work out each corresponding value of y.

3. The **only** column that changes sign is the x term (middle) column.
   If the given values of x contain 0, then the x term column will make one sign change, either from + to − or from − to +, where $x = 0$.

4. The other two columns **never** change sign. They remain either all +'s or all −'s.
   These columns keep the sign given in the question.

Note: Decide where to draw the x- and y-axes by looking at the table to see what the largest and smallest values of x and y are. In general, the units on the x-axis are larger than the units on the y-axis. Try to make sure that the graph extends almost the whole width and length of the page.

Example ▼

Graph the function $g : x \to 5 + 3x - 2x^2$, in the domain $-2 \leq x \leq 3$, $\quad x \in \mathbf{R}$.

Solution:

Let $y = g(x) \Rightarrow y = 5 + 3x - 2x^2$

x	$-2x^2 + 3x + 5$	y
−2	−8 − 6 + 5	−9
−1	−2 − 3 + 5	0
0	−0 + 0 + 5	5
1	−2 + 3 + 5	6
2	−8 + 6 + 5	3
3	−18 + 9 + 5	−4

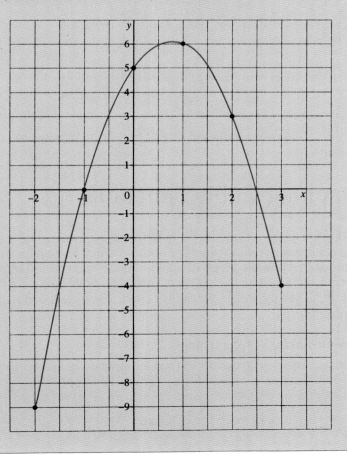

Exercise 7.2

Graph each of the following functions in the given domain:

1. $g : x \rightarrow x^2 - 2x - 8$, in the domain $-3 \leqslant x \leqslant 5$, $x \in \mathbf{R}$.

2. $f : x \rightarrow x^2 + 2x - 3$, in the domain $-4 \leqslant x \leqslant 2$, $x \in \mathbf{R}$.

3. $f : x \rightarrow 2x - x^2$, in the domain $-2 \leqslant x \leqslant 4$, $x \in \mathbf{R}$.

4. $g : x \rightarrow 2x^2 + 3x - 2$, in the domain $-3 \leqslant x \leqslant 2$, $x \in \mathbf{R}$.

5. $f : x \rightarrow 2x^2 - x - 4$, in the domain $-2 \leqslant x \leqslant 3$, $x \in \mathbf{R}$.

6. $g : x \rightarrow 3 + 5x - 2x^2$, in the domain $-1 \leqslant x \leqslant 4$, $x \in \mathbf{R}$.

7. On the same axes and scales, graph the functions:
   $f : x \rightarrow x^2 - 2x - 4$ $g : x \rightarrow 2x + 1$, in the domain $-3 \leqslant x \leqslant 5$, $x \in \mathbf{R}$.

8. On the same axes and scales, graph the functions:
   $f : x \rightarrow 5 + 2x - x^2$, $g : x \rightarrow 2 - x$, in the domain $-2 \leqslant x \leqslant 4$, $x \in \mathbf{R}$.

9. On the same axes and scales, graph the functions:
   $f : x \rightarrow 2x^2 - 3x - 8$, $g : x \rightarrow 3x - 2$, in the domain $-2 \leqslant x \leqslant 4$, $x \in \mathbf{R}$.

10. On the same axes and scales, graph the functions:
    $f : x \rightarrow 5 - x - 2x^2$, $g : x \rightarrow 1 - 3x$, in the domain $-2 \leqslant x \leqslant 3$, $x \in \mathbf{R}$.
    (i) From your graph, write down the coordinates of the points of intersection of f and g.
    (ii) Verify the x values in part (i) by solving the equation $g(x) = f(x)$.

Graphing Cubic Functions

A **cubic** function is usually given in the form $f : x \rightarrow ax^3 + bx^2 + cx + d$, $a \neq 0$ and a, b, c and d are constants. For example, $f : x \rightarrow x^3 - 2x^2 - 6x + 5$. As with graphing quadratic functions, quite a few points are needed to plot the graph of a cubic function. In the question, you will always be given a set of inputs, x, called the domain. With these inputs, a table is used to find the corresponding set of outputs, y or $f(x)$, called the range. When the table is completed, plot the points and join them with a '**smooth curve**'.

Notes on making out the table:

1. Work out each column separately, i.e. all the x^3 terms first, then all the x^2 terms, then all the x terms and finally the constant term. (Watch for patterns in the numbers.)
2. Work out each corresponding value of y.

GRAPHING FUNCTIONS

3. Only **two** columns may change sign, the x^3 and x term columns. If the given values of x contain 0, then the x^3 and x term columns will make **one** sign change, either from + to − or − to +, where $x = 0$. So the only two signs to work out are the ones at the beginning of the x^3 and x term columns, which are changed at $x = 0$. The simplest way to find the first sign in these two columns is to multiply the sign of the first given value of x by the sign of the coefficient of x^3 and x.

4. The other two columns **never** change sign. They remain all +'s or all −'s. These columns keep the sign given in the question.

Note: Decide where to draw the x- and y-axes by looking at the table to see what the largest and smallest values of x and y are. In general, the units on the x-axis are larger than the units on the y-axis. Try to make sure that the graph extends almost the whole width and length of the page.

Example ▼

Graph the function $f : x \to x^3 + 3x^2 - x - 3$ in the domain $-4 \leqslant x \leqslant 2$, $x \in \mathbf{R}$.

Solution:

Let $y = f(x) \Rightarrow y = x^3 + 3x^2 - x - 3$.

x	$x^3 + 3x^2 - x - 3$	y
−4	−64 + 48 + 4 − 3	−15
−3	−27 + 27 + 3 − 3	0
−2	−8 + 12 + 2 − 3	3
−1	−1 + 3 + 1 − 3	0
0	0 + 0 + 0 − 3	−3
1	1 + 3 − 1 − 3	0
2	8 + 12 − 2 − 3	15

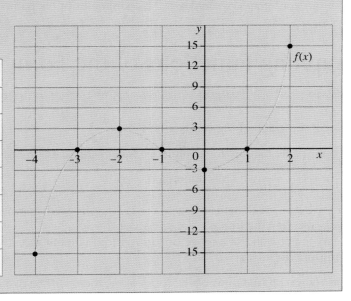

Exercise 7.3 ▼

Graph each of the following functions in the given domain:

1. $f : x \to x^3 + x^2 - 2x - 1$ in the domain $-3 \leqslant x \leqslant 2$, $x \in \mathbf{R}$.
2. $g : x \to x^3 - 4x^2 + x + 6$ in the domain $-2 \leqslant x \leqslant 4$, $x \in \mathbf{R}$.
3. $f : x \to 2x^3 - 3x^2 - 6x + 2$ in the domain $-2 \leqslant x \leqslant 3$, $x \in \mathbf{R}$.
4. $g : x \to 2x^3 - 2x^2 - 3x + 10$ in the domain $-2 \leqslant x \leqslant 2$, $x \in \mathbf{R}$.
5. $f : x \to 6 + 6x - x^2 - x^3$ in the domain $-3 \leqslant x \leqslant 3$, $x \in \mathbf{R}$.

6. $g: x \to 6 + 5x - 2x^2 - x^3$ in the domain $-4 \leq x \leq 3$, $x \in \mathbf{R}$.

7. $f: x \to 4 + 8x + x^2 - 2x^3$ in the domain $-2 \leq x \leq 3$, $x \in \mathbf{R}$.

8. $g: x \to 2x - x^2 - x^3$ in the domain $-3 \leq x \leq 2$, $x \in \mathbf{R}$.

9. $f: x \to x^3 - 3x^2 - 2$ in the domain $-2 \leq x \leq 4$, $x \in \mathbf{R}$.

10. $g: x \to 3 + 5x - x^3$ in the domain $-3 \leq x \leq 3$, $x \in \mathbf{R}$.

11. On the same axes and scales, graph the functions:
$f: x \to x^3 - 2x^2 - 4x + 1$ and $g: x \to 3x + 1$, in the domain $-2 \leq x \leq 4$, $x \in \mathbf{R}$.

12. On the same axes and scales, graph the functions:
$f: x \to x^3 + 3x^2 - x - 3$ and $g: x \to 3 - 2x - x^2$, in the domain $-4 \leq x \leq 2$, $x \in \mathbf{R}$.
From your graph, write down the coordinates of the points of intersection of f and g.

13. On the same axes and scales, graph the functions:
$f: x \to x^3 - 2x^2 - 6x + 2$, $g: x \to 2x^2 - 7x - 4$, $h: x \to 2 - 3x$ in the domain $-2 \leq x \leq 4$, $x \in \mathbf{R}$.
From your graph, write down the coordinates of the points of intersection of f, g and h.

Graphing Functions of the Form $\dfrac{1}{x+a}$

Consider the graph of the function $f: x \to \dfrac{1}{x-2}$.

It is impossible to draw the complete graph of the function because when $x = 2$,
$\dfrac{1}{x-2} = \dfrac{1}{2-2} = \dfrac{1}{0}$ which is undefined.

There will be a break in the curve at $x = 2$. The curve will approach the line $x = 2$, getting closer without actually meeting the line $x = 2$ at any stage.
If a curve approaches a line, getting closer to it without actually meeting the line, the line is called an **'asymptote'** to the curve.

When you graph functions of the form $\dfrac{1}{x+a}$, there will be two asymptotes to the curve.

1. $x = -a$ (vertical asymptote)
2. $y = 0$ (the x-axis, horizontal asymptote)

On the right is a rough graph of the function $f(x) = \dfrac{1}{x-2}$.

Notice that the curve **never** meets the line $x = 2$ and **never** meets the x-axis.

All functions of the form $\dfrac{1}{x+a}$ will have a similar shape to the graph on the right.

This curve is called a **'rectangular hyperbola'**.
From the graph we can see that as x increases, y decreases, and vice versa.

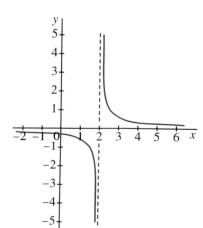

GRAPHING FUNCTIONS

Method for Graphing Functions of the Form $\frac{1}{x+a}$

1. Let the expression on the bottom = 0, to find its vertical asymptote.
2. Represent this vertical asymptote with a broken line.
3. $y = 0$, the x-axis, will always be the horizontal asymptote.
4. Use a table to find suitable coordinates on the curve.
   Make sure to take some values of x very close to the vertical asymptote.
5. Join these points with a smooth curve.

Example ▼

Graph the function $f: x \rightarrow \dfrac{1}{x+1}$ in the domain $-4 \leqslant x \leqslant 2$, $x \in \mathbf{R}$, $x \neq -1$.

Solution:
Expression on the bottom = 0
$$x + 1 = 0$$
$$x = -1 \quad \text{[this is the vertical asymptote (broken line on the graph)]}$$
The x-axis is the horizontal asymptote.

Let $y = f(x) \Rightarrow y = \dfrac{1}{x+1}$

x	$\dfrac{1}{x+1}$	y
-4	$\dfrac{1}{-4+1}$	$-\dfrac{1}{3}$
-3	$\dfrac{1}{-3+1}$	$-\dfrac{1}{2}$
-2	$\dfrac{1}{-2+1}$	-1
$-1\tfrac{1}{2}$	$\dfrac{1}{-1\tfrac{1}{2}+1}$	-2
$-\tfrac{1}{2}$	$\dfrac{1}{-\tfrac{1}{2}+1}$	2
0	$\dfrac{1}{0+1}$	1
1	$\dfrac{1}{1+1}$	$\dfrac{1}{2}$
2	$\dfrac{1}{2+1}$	$\dfrac{1}{3}$

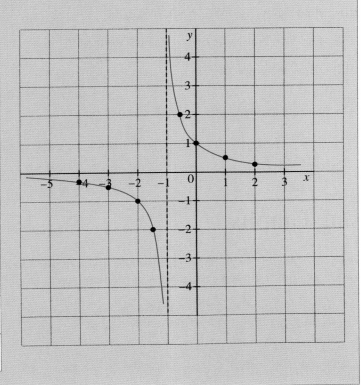

243

Exercise 7.4

Graph each of the following functions in the given domain:

1. $f: x \to \dfrac{1}{x-1}$, in the domain $-2 \leq x \leq 4$, $x \in \mathbf{R}$, $x \neq 1$.

2. $g: x \to \dfrac{1}{x-2}$, in the domain $-1 \leq x \leq 5$, $x \in \mathbf{R}$, $x \neq 2$.

3. $f: x \to \dfrac{1}{x+3}$, in the domain $-7 \leq x \leq 1$, $x \in \mathbf{R}$, $x \neq -3$.

4. $g: x \to \dfrac{1}{x-4}$, in the domain $0 \leq x \leq 8$, $x \in \mathbf{R}$, $x \neq 4$.

5. $f: x \to \dfrac{1}{x+2}$, in the domain $-6 \leq x \leq 2$, $x \in \mathbf{R}$, $x \neq -2$.

6. $g: x \to \dfrac{1}{x-3}$, in the domain $-1 \leq x \leq 7$, $x \in \mathbf{R}$, $x \neq 3$.

7. $f: x \to \dfrac{1}{x}$, in the domain $-4 \leq x \leq 4$, $x \in \mathbf{R}$, $x \neq 0$.

8. Let $f(x) = \dfrac{1}{x-5}$.

   (i) Find $f(1), f(3), f(4\tfrac{1}{2}), f(5\tfrac{1}{2}), f(6), f(8)$.

   (ii) For what real value of x is $f(x)$ not defined?

   (iii) Draw the graph of $f(x) = \dfrac{1}{x-5}$ in the domain $1 \leq x \leq 9$, $x \in \mathbf{R}$.

9. On the same axes and scales, graph the functions:

   $f: x \to \dfrac{1}{x+1}$ and $g: x \to x+1$, in the domain $-5 \leq x \leq 3$, $x \in \mathbf{R}$.

Using Graphs

Once we have drawn the graph, we are usually asked to use it to answer some questions. Below are examples of the general types of problem where graphs are used.

Notes: 1. $y = f(x)$, so $f(x)$ can be replaced by y.
2. In general, if given x find y, and vice versa.

GRAPHING FUNCTIONS

Examples of the main types of problem, once the graph is drawn:

1. **Find the values of x for which $f(x) = 0$.**
   This question is asking:
   'Where does the curve meet the x-axis?'

   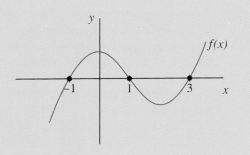

   Solution:
   Write down the values of x where the graph meets the x-axis.
   From the graph: $x = -1$ or $x = 1$ or $x = 3$

2. **Find the values of x for which $f(x) = 6$.**
   This question is asking:
   'When $y = 6$, what are the values of x?'

   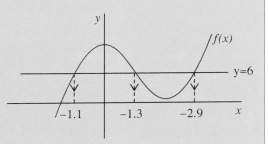

   Solution:
   Draw the line $y = 6$. Where this line meets the curve draw broken perpendicular lines onto the x-axis.
   Write down the values of x where these broken lines meet the x-axis.
   From the graph:
   When $y = 6$, $\quad x = -1.1$ or $x = 1.3$ or $x = 2.9$

3. **Find the value of $f(-1.6)$.**
   This question is asking:
   'When $x = -1.6$, what is the value of y?'

   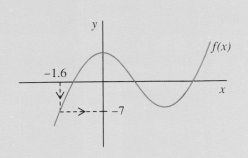

   Solution:
   From $x = -1.6$ on the x-axis draw a broken perpendicular line to meet the curve. From this draw a broken horizontal line to meet the y-axis. Write down the value of y where this line meets the y-axis.
   From the graph:
   $f(-1.6) = -7$

4. Local maximum and minimum points or the local maximum and minimum values

Often we are asked to find the local maximum and minimum points or the local maximum and minimum values. Consider the graph on the right. The local maximum and minimum points are where the graph turns, $(1, 4)$ and $(4, -2)$, respectively. The local maximum and minimum values are found by drawing a line from the turning points to the y-axis and reading the values where these lines meet the y-axis. The maximum and minimum values are 4 and -2, respectively.

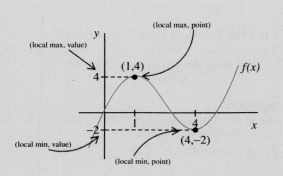

5. Increasing and decreasing

Graphs are read from left to right.

Increasing: $f(x)$ is increasing where the graph is **rising** as we go from left to right.

Decreasing: $f(x)$ is decreasing where the graph is **falling** as we go from left to right.

Find the values of x for which:

(i) $f(x)$ is increasing

(ii) $f(x)$ is decreasing.

Solution:

(i) $f(x)$ increasing, graph rising from left to right.
The values of x are:
$-4 \leqslant x < -2$ and $1 < x \leqslant 3$

(ii) $f(x)$ decreasing, graph falling from left to right.
The values of x are: $-2 < x < 1$

Note: At $x = -2$ and $x = 1$, the graph is neither increasing nor decreasing.

6. Positive and negative

Positive, $f(x) > 0$: Where the graph is **above** the x-axis.

Negative, $f(x) < 0$: Where the graph is **below** the x-axis.

Find the values of x for which: (i) $f(x) > 0$ (ii) $f(x) < 0$.

Solution:

(i) $f(x) > 0$, curve **above** the x-axis.
The values of x are:
$-1 < x < 1$ and $3 < x \leqslant 4$

(ii) $f(x) < 0$, curve **below** the x-axis.
The values of x are:
$-2 \leqslant x < -1$ and $1 < x < 3$

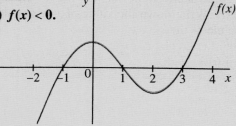

Note: If the question uses $f(x) \geqslant 0$ or $f(x) \leqslant 0$, then the values of x where the graph meets the x-axis must also be included.

GRAPHING FUNCTIONS

7. Two functions graphed on the same axes and scales
The diagram shows the graph of two functions: $f(x)$, a curve, and $g(x)$, a line.

Find the values of x for which:
(i) $f(x) = g(x)$ (ii) $f(x) \leqslant g(x)$ (iii) $f(x) \geqslant g(x)$

Solution:
(i) $f(x) = g(x)$
(curve = line)
The values of x are: 0.4, 1.4 and 3.1

(ii) $f(x) \leqslant g(x)$
(curve equal to and below the line)
The values of x are:
$-1 \leqslant x \leqslant 0.4$ and $1.4 \leqslant x \leqslant 3.1$

(iii) $f(x) \geqslant g(x)$
(curve equal to and above the line)
The values of x are:
$0.4 \leqslant x \leqslant 1.4$ and $3.1 \leqslant x \leqslant 4$

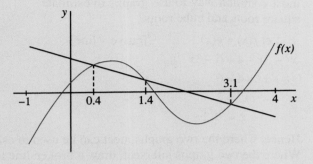

Note: If the question uses $f(x) < g(x)$ or $f(x) > g(x)$, then the values of x where the graphs meet (0.4, 1.4 and 3.1) are **not** included in the solution.

8. Graph above or below a constant value (an inequality)

Find the values of x for which: (i) $f(x) \geqslant 2$ (ii) $f(x) \leqslant 2$.
These questions are asking:
'What are the values of x for which the curve, $f(x)$, is (i) 2 or above (ii) 2 or below?'

Solution:
Draw the line $y = 2$.
Write down the values of x for which the curve is:
(i) on or above the line $y = 2$
(ii) on or below the line $y = 2$.

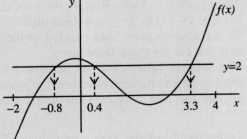

(i) $f(x) \geqslant 2$, curve on or above the line $y = 2$.
The values of x are: $-0.8 \leqslant x \leqslant 0.4$ and $3.3 \leqslant x \leqslant 4$
(ii) $f(x) \leqslant 2$, curve on or below the line $y = 2$.
The values of x are: $-2 \leqslant x \leqslant -0.8$ and $0.4 \leqslant x \leqslant 3.3$

Note: If the question uses $f(x) > 2$ or $f(x) < 2$, then the values of x where the curve meets the line $y = 2$ (−0.8, 0.4 and 3.3) are **not** included in the solution.

9. **Using two graphs to estimate square roots and cube roots**
The diagram shows graphs of the functions:
$f : x \rightarrow x^3 - 3x + 4$ and $g : x \rightarrow 6 - 3x$, in the domain $-2 \leq x \leq 3$.
Show how the graphs may be used to estimate the value of $\sqrt[3]{2}$.

Solution:

The values of x where two graphs meet is the most common way to use graphs to estimate square roots and cube roots.

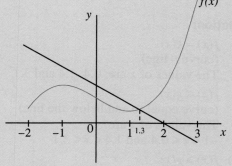

Let $f(x) = g(x)$ [curve = line]
$$x^3 - 3x + 4 = 6 - 3x$$
$$x^3 + 4 = 6$$
$$x^3 = 2$$
$$x = \sqrt[3]{2}$$

Hence, where the two graphs meet can be used to estimate the value of $\sqrt[3]{2}$.
Where the two graphs intersect, draw a broken line to meet the x-axis.
This line meets the x-axis at 1.3.
Thus, using the graphs, we estimate the value of $\sqrt[3]{2}$ to be 1.3.

Note: A calculator gives $\sqrt[3]{2} = 1.25992104989$

10. **Number of times a graph meets the x-axis**
The number of times a graph meets the x-axis gives the number of roots of its equation. Often we need to find a range of values of a constant, which shifts a graph up or down, giving a graph a certain number of roots, e.g.:

For what values of k does the equation $f(x) = k$ have three roots?

Solution:

The equation $f(x) = k$ will have three roots if the line $y = k$ cuts the graph three times.
So we have to draw lines parallel to the x-axis that cut the graph three times.
The range of values of k will be in between the lowest and highest values on the y-axis, so that the line $y = k$ cuts the graph three times.

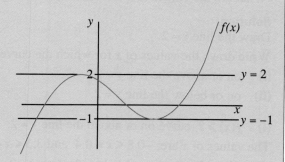

Question: Find the range of values of k for which $f(x) = k$ has three roots.

Solution:
The range of values of k is found by finding the range of the equations of the lines, parallel to the x-axis, which cut the graph three times.

Any lines drawn parallel to the y-axis between $y = -1$ and $y = 2$, will cut the graph three times
∴ k will lie between -1 and 2
∴ $f(x) = k$ will have three roots for $-1 < k < 2$.

GRAPHING FUNCTIONS

Example

Graph the function $f: x \to 2 - 9x + 6x^2 - x^3$, in the domain $-1 \leq x \leq 5$, $x \in \mathbf{R}$.
Use your graph to estimate:

(i)   the values of x for which $f(x) = 0$

(ii) $f(-0.5)$

(iii) the values of x for which $f(x) > 0$ and increasing

(iv) the range of real values of k for which $f(x) = k$ has more than one solution.

Solution:

Let $y = f(x) \Rightarrow y = 2 - 9x + 6x^2 - x^3$

x	$-x^3$	$+6x^2$	$+9x$	$+2$	y
-1	$+1$	$+6$	$+9$	$+2$	18
0	$+0$	$+0$	$+0$	$+2$	2
1	-1	$+6$	-9	$+2$	-2
2	-8	$+24$	-18	$+2$	0
3	-27	$+54$	-27	$+2$	2
4	-64	$+96$	-36	$+2$	-2
5	-125	$+150$	-45	$+2$	-18

(i) **Estimate the values of x for which $f(x) = 0$.**
This question is asking, 'Where does the curve meet the x-axis?'
The curve meets the x-axis at 0.3, 2 and 3.7.
Therefore, the values of x for which $f(x) = 0$, are 0.3, 2 and 3.7.

Note: 'Find the values of x for which $2 - 9x + 6x^2 - x^3 = 0$' is another way of asking the same question.

(ii) **Estimate the value of $f(-0.5)$.**
This question is asking, 'When $x = -0.5$, what is the value of y?'
From $x = -0.5$ on the x-axis draw a broken perpendicular line to meet the curve.
From this draw a broken horizontal line to meet the y-axis.
This line meets the y-axis at 7.9.
Therefore $f(-0.5) = 7.9$.

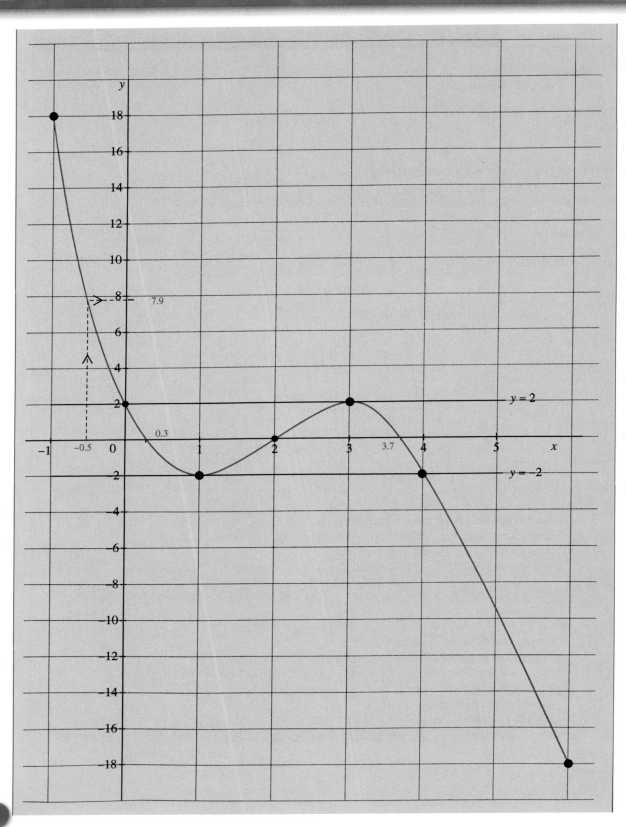

(iii) **Estimate the values of x for which $f(x) > 0$ and is increasing.**
This question is asking,
'Where is the curve above the x-axis and increasing as we go from left to right?'
From the graph, the curve is above the x-axis and increasing between 2 and 3.
Therefore, the values of x for which $f(x) > 0$ and increasing are $2 < x < 3$.

Note: $x = 2$ is not included because at $x = 2$, $f(x) = 0$ and we are given $f(x) > 0$.
$x = 3$ is not included because at $x = 3$ there is a turning point and $f(x)$ is not increasing.

(iv) **Estimate the range of real values of k for which $f(x) = k$ has more than one solution.**
The range of values of k is found by drawing lines parallel to the x-axis that meet the graph more than once.
Any lines drawn parallel to the x-axis between -2 and 2 will meet the curve more than once.
Therefore, k will lie between -2 and 2, including -2 and 2.
Therefore, $f(x) = k$ will have more than one solution when $-2 \leqslant k \leqslant 2$.

Example ▼

On the same axis and scales, graph the functions:
$f : x \rightarrow x^3 - 6x + 2$ and $g : x \rightarrow 2x - 2$, in the domain $-3 \leqslant x \leqslant 3$, $x \in \mathbf{R}$.
Use your graph to estimate:

(i) the values of x for which $f(x) = 2$.
Hence, estimate $\sqrt{6}$, giving a reason for your answer.

(ii) the range of values of x for which $g(x) > f(x)$

(iii) the roots of the equation $x^3 - 8x + 4 = 0$.

Solution:

Let $y = f(x) \Rightarrow y = x^3 - 6x + 2$

x	$x^3 - 6x + 2$	y
-3	$-27 + 18 + 2$	-7
-2	$-8 + 12 + 2$	6
-1	$-1 + 6 + 2$	7
0	$+0 + 0 + 2$	2
1	$+1 - 6 + 2$	-3
2	$+8 - 12 + 2$	-2
3	$+27 - 18 + 2$	11

Let $y = g(x) \Rightarrow y = 2x - 2$

x	$2x - 2$	y
-3	$-6 - 2$	-8
3	$6 - 2$	4

(using the smallest and largest values of x)

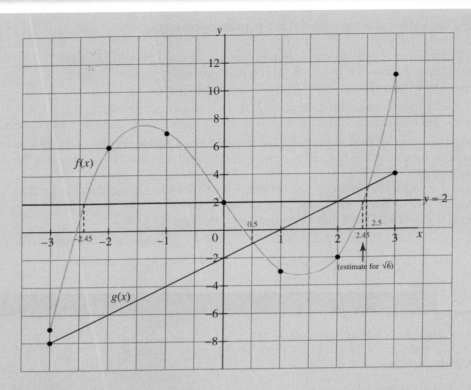

(i) Estimate the values of x for which $f(x) = 2$.

This question is asking, 'When $y = 2$, what are the values of x?'
Draw the line $y = 2$.
Where this line meets the curve draw broken perpendicular lines to meet the x-axis.
These lines meet the x-axis at -2.45, 0 and 2.45
Therefore, the values of x for which $f(x) = 2$ are -2.45, 0 and 2.45.

Note: 'Estimate the values of x for which $x^3 - 6x + 2 = 2$' is another way of asking the same question.

Use $f(x) = 2$ to estimate $\sqrt{6}$.

Let
$$f(x) = 2$$
$$x^3 - 6x + 2 = 2 \quad (f(x) = x^3 - 6x + 2)$$
$$x^3 - 6x = 0 \quad \text{(subtract 2 from both sides)}$$
$$x(x^2 - 6) = 0 \quad \text{(factorise the left-hand side)}$$
$$x = 0 \quad \text{or} \quad x^2 - 6 = 0 \quad \text{(let each factor} = 0\text{)}$$
$$x = 0 \quad \text{or} \quad x^2 = 6$$
$$x = 0 \quad \text{or} \quad x = \pm\sqrt{6}$$

Therefore, when $f(x) = 2$, one of the roots is $\sqrt{6}$.
Thus, the line $y = 2$ can be used to estimate $\sqrt{6}$.
One of the values of x where the line $y = 2$ intersects the curve will give an estimate for $\sqrt{6}$.
From the graph, $\sqrt{6}$ is approximately 2.45.

(ii) **Estimate the range of values of x for which $g(x) > f(x)$.**
$f(x)$ is the curve, $g(x)$ is the line.
This question is asking, 'What are the values of x where the line is **above** the curve?'
First work out the values of x where the curve and line intersect.
Where the curve and line meet, draw broken perpendicular lines to meet the x-axis.
These lines meet the x-axis at 0.5 and 2.5.
Therefore, the values of x for which $f(x) = g(x)$ are 0.5 and 2.5.
From the graphs, the line is above the curve between 0.5 and 2.5.
Therefore, the values of x for which $g(x) > f(x)$ are $0.5 < x < 2.5$.

(iii) **Estimate the roots of the equation $x^3 - 8x + 4 = 0$.**
We have to use the graph to solve this equation.

$$x^3 - 8x + 4 = 0$$
$$(x^3 - 6x + 2) - 2x + 2 = 0 \qquad \text{(rearranging in terms of } f(x)\text{)}$$
$$f(x) - 2x + 2 = 0 \qquad (f(x) = x^3 - 6x + 2)$$
$$f(x) = 2x - 2$$
$$f(x) = g(x) \qquad (g(x) = 2x - 2)$$

$f(x) = g(x)$ is where the curve and the line meet.
Thus, the values of x where the curve and the line meet are solutions of the equation $x^3 - 8x + 4 = 0$.
From the graph, the values of x where the curve and line meet are 0.5 and 2.5.
Therefore, the roots of the equation $x^3 - 8x + 4 = 0$ are 0.5 and 2.5.

Note: There is another root, -3.1, of the equation $x^3 - 8x + 4 = 0$, but it is outside the given domain, $-3 \leq x \leq 3$, and thus we cannot use the graphs to find this root.

Exercise 7.5

1. Below is a graph of the function $f: x \rightarrow x^3 - 3x^2 + 4$, in the domain $-2 \leq x \leq 4$, $\quad x \in \mathbf{R}$.

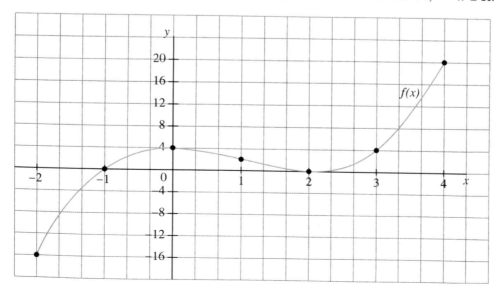

Use your graph to:
(i) find the values of x for which $f(x) = 0$
(ii) find the coordinates of the local maximum and minimum points of f
(iii) find the local maximum and minimum values
(iv) estimate the value of $f(3.5)$
(v) find the values of x for which $f(x)$ is decreasing
(vi) find the values of x for which $f(x)$ is increasing
(vii) find the values of x for which $f(x) < 0$ and increasing
(viii) find the values of k for which $f(x) = k$ has three solutions
(ix) estimate the values of x for which   (a) $f(x) = 2$ (b) $f(x) \geqslant 2$.

2. Graph the function $f: x \rightarrow x^3 + 4x^2 + x - 6$ in the domain $-4 \leqslant x \leqslant 2$, $x \in \mathbf{R}$.
Use your graph to find the values of x for which:
(i) $f(x) = 0$ (ii) $f(x) \geqslant 0$ (iii) $f(x) \leqslant 0$ and $x \leqslant 0$.
Estimate the value of $f(1.4)$.

3. Graph the function $f: x \rightarrow x^3 - 2x^2 - 5x + 6$ in the domain $-3 \leqslant x \leqslant 4$, $x \in \mathbf{R}$.
Use your graph to:
(i) find the values of x for which $f(x) = 0$
(ii) find the values of x for which: (a) $f(x) > 0$ (b) $f(x) < 0$.
(iii) estimate the values of x for which $f(x) = 5$.

4. Let $f(x) = 2x^3 - 3x^2 - 12x + 4$ for $x \in \mathbf{R}$.
(i) Complete the table:

x	-2.5	-2	-1	0	1	2	3	3.5
$f(x)$	-16							11

(ii) Draw the graph of $f(x)$ in the domain $-2.5 \leqslant x \leqslant 3.5$, $x \in \mathbf{R}$.
(iii) Write down the coordinates of the local maximum and the local minimum points.
(iv) Use your graph to find the values of x for which $f(x)$ is:
  (a) increasing   (b) decreasing   (c) positive, increasing and $x < 0$.
(v) Estimate the values of x for which $f(x) = 0$.

5. The function $f: x \rightarrow 2x^3 - 5x^2 - 2x + 5$ is defined on the domain $-2 \leqslant x \leqslant 3$ for $x \in \mathbf{R}$.
Verify that $f(-2) = -27$ and $f(-1) = 0$ and draw the graph of f.
Use your graph to solve the equation $2x(x - 1)(x + 2) = 7x^2 - 2x - 5$.
Use your graph to estimate:
(i) the value of $f(1.75)$
(ii) the values of x for which $2x^3 - 5x^2 - 2x = 0$.

6. Graph the function $g: x \rightarrow x^3 - 4x^2 + 5$, in the domain $-2 \leqslant x \leqslant 4$, $x \in \mathbf{R}$.
(i) Write down the coordinates of the local maximum point.
(ii) Estimate the values of x for which $f(x) = 0$.
(iii) Use your graph to solve the equation $x^3 - 4x^2 + 2 = 0$.
Find the range of values of h, in the given domain, for which $f(x) = h$ has only one solution.

7. Graph the function $f: x \rightarrow 12x + 3x^2 - 2x^3$, in the domain $-2 \leq x \leq 3.5$, $x \in \mathbf{R}$.
   Use your graph to estimate the values of x for which:
   (i) $f(x) = 0$ (ii) $f(x) = 2$ (iii) $f(x) \leq 0$ and $x > 0$.
   (iv) Write down the coordinates of the local maximum and local minimum points.
   (v) Find the range of values of k for which $f(x) = k$ has:
       (a) three real roots   (b) two real roots.

8. On the same axes and scales, graph the functions:
   $f: x \rightarrow x^3 - 3x^2 - 4x + 12$, $g: x \rightarrow 9 - 3x$, in the domain $-2 \leq x \leq 4$, $x \in \mathbf{R}$.
   Use your graphs to find the values of x for which:
   (i) $f(x) = 0$ (ii) $f(x) = 12$ (iii) $f(x) = g(x)$ (iv) $f(x) \geq g(x)$ (v) $f(x) \leq g(x)$.

9. On the same axes and scales, graph the functions:
   $f: x \rightarrow x^3 - 2x^2 - 6x + 4$ and $g: x \rightarrow 2x - 5$, in the domain $-2 \leq x \leq 4$, $x \in \mathbf{R}$.
   Use your graphs to find:
   (i) the two values of x for which $f(x) = g(x)$
   (ii) the range of values of x for which $f(x) \leq g(x)$
   (iii) the range of values of x for which $f(x) \geq g(x)$.

10. On the same axes and scales, graph the functions:
    $f: x \rightarrow 2x^3 - 3x^2 - 3x + 2$, $g: x \rightarrow 3x + 2$, in the domain $-2 \leq x \leq 3$, $x \in \mathbf{R}$.
    Use your graphs to estimate the values of x for which:
    (i) $f(x) = 0$ (ii) $2x^3 - 3x^2 - 6x = 0$ (iii) $2x^3 - 3x^2 - 6x \geq 0$.

11. Graph the function $f: x \rightarrow 2 + 3x^2 - x^3$, in the domain $-1 \leq x \leq 4$, $x \in \mathbf{R}$.
    (i) Write down the coordinates of the local maximum and local minimum points.
    (ii) Estimate the value of x for which $f(x) = 0$.
    (iii) Write the equation $-2x + 3x^2 - x^3 = 0$ in the form $2 + 3x^2 - x^3 = ax + a$.
    (iv) Using the value of a from part (iii), and on the same axes and scales, graph the function
         $g: x \rightarrow ax + a$, in the domain $-1 \leq x \leq 4$, $x \in \mathbf{R}$.
    (v) Hence, use your graphs to find the solutions of the equation $-2x + 3x^2 - x^3 = 0$.

12. Let $f(x) = x^3 + 2x^2 - 7x - 2$ for $x \in \mathbf{R}$.
    (i) Complete the following table:

x	-4	-3	-2	-1	0	1	2	3
$f(x)$		10				-6		

    (ii) Draw the graph of f.
    (iii) Estimate the values of x for which $f(x) = 0$.
    (iv) Use the graph to find the least value of $f(x)$ in $0 \leq x \leq 3$.
    (v) If $g: x \rightarrow f(x) + k$, find the value of k when the x-axis is a tangent to the graph of g in $0 \leq x \leq 3$.
    (vi) By drawing an appropriate line, use both graphs to solve $x^3 + 2x^2 - 5x - 6 = 0$.

13. (i) Verify that $x = 2$ is a root of the equation $x^3 - 5x^2 + 2x + 8 = 0$ and find the other two roots.
    Graph the function $f: x \to x^3 - 5x^2 + 2x + 8$ in the domain $-2 \leq x \leq 5$, $x \in \mathbf{R}$.
    (ii) Use your graph to find the values of x for which $f(x) = 0$ and verify your answers to part (i).
    (iii) Write the equation $x^3 - 5x^2 + 4x = 0$ in the form $x^3 - 5x^2 + 2x + 8 = a - bx$.
    (iv) Using the values of a and b from part (iii), and on the same axes and scales, graph the function $g: x \to a - bx$, in the domain $-2 \leq x \leq 5$, $x \in \mathbf{R}$.
    (v) Use your graphs to find the solutions of the equation $x^3 - 5x^2 + 4x = 0$.
    (vi) Hence, or otherwise, write down the values of x for which $x^3 - 5x^2 + 4x \geq 0$, in the domain $-2 \leq x \leq 5$, $x \in \mathbf{R}$.

14. Factorise $x^3 - 4x$ and hence solve the equation $x^3 - 4x = 0$.
    Graph the function $f: x \to x^3 - 5x + 2$, in the domain $-3 \leq x \leq 3$, $x \in \mathbf{R}$.
    (i) Estimate from your graph the three values of x for which $f(x) = 0$.
    (ii) Estimate from your graph the three values of x for which $f(x) = 2$ and hence, estimate $\sqrt{5}$, giving a reason for your answer.

15. Graph the function $f: x \to x^3 - 3x + 2$ in the domain $-3 \leq x \leq 3$, $x \in \mathbf{R}$.
    Use your graph to find the following:
    (i) the local maximum and minimum values.
    (ii) the coordinates of the local maximum and minimum points.
    (iii) the range of values of x for which $f(x)$ is:
        (a) decreasing (b) negative and increasing (iii) positive and increasing.
    (iv) the range of real values of k for which the equation $x^3 - 3x + 2 = k$ has three real and distinct roots.
    Estimate from your graph the values of x for which $f(x) = 2$.
    Hence, estimate $\sqrt{3}$, giving a reason for your answer.

16. Let $f(x) = x^3 + x^2 + x - 2$ for $x \in \mathbf{R}$.
    (i) Complete the table:

x	−2	−1	0	1	2
$f(x)$	−8				

    (ii) Draw the graph of $f(x)$ in the domain $-2 \leq x \leq 2$.
    (iii) Use the same axes and scales, and in the same domain, graph the function $g: x \to x^2 + x - 6$, for $x \in \mathbf{R}$.
    (iv) Use your graphs to estimate $\sqrt[3]{-4}$.

17. On the same axes and scales, graph the functions:
    $g: x \to \dfrac{1}{x-1}$, $h: x \to x + 1$ in the domain $-2 \leq x \leq 4$, $x \in \mathbf{R}$.
    Show how your graphs may be used to estimate the value of $\pm\sqrt{2}$.

GRAPHING FUNCTIONS

Periodic Functions

A function whose graph repeats itself at regular intervals is called '**periodic**'.
For example, the graph below is a graph of a periodic function, $y = f(x)$.

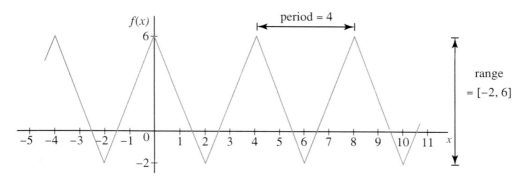

The '**period**' is the horizontal width it takes for a graph to repeat itself.
In this case the period = 4 (graph repeats itself after a distance of 4).
The '**range**' is the interval from the least value of y to the greatest value of y.
In this case the range = $[-2, 6]$ (lower value is written first).
A feature of the graphs of periodic functions is that we can add, or subtract, the period, or integer multiples of the period, to the value of $f(x)$ and the value of the function is unchanged.
For example, using the graph above:

$$f(-4) = f(0) = f(4) = f(8) = 6 \qquad [f(-4) = f(-4+4) = f(-4+8) = f(-4+12) \ldots]$$

$$f(10) = f(6) = f(2) = f(-2) = -2 \qquad [f(10) = f(10-4) = f(10-8) = f(10-12) \ldots]$$

On our course we will be given the graph of a periodic function on scaled and labelled axes.
The period will be a positive whole number and the range will be a closed interval $[a, b]$, where a and b are whole numbers, positive or negative.

Part of the graph of the periodic function $y = f(x)$ is shown below.
State the period and range of the function.

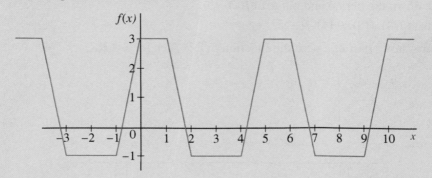

Evaluate: $f(1), f(4), f(20), f(23.5)$.

Solution:

Basic building block

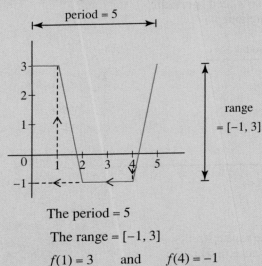

The period = 5

The range = [−1, 3]

$f(1) = 3$ and $f(4) = −1$

Basic building block

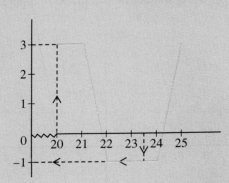

$f(20) = 3$ and $f(23.5) = −1$

Alternatively, by repeatedly subtracting the period 5:

$f(20) = f(15) = f(10) = f(5) = 3$

$f(23.5) = f(18.5) = f(13.5) = f(8.5) = f(3.5) = −1$

Exercise 7.6

1. The diagram shows part of a periodic function $f : x \to f(x)$, for $x \in \mathbf{R}$.

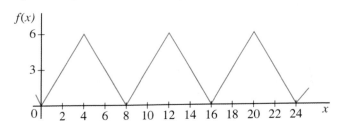

(i) Write down the period and range of $f(x)$.

(ii) Evaluate $f(4), f(8), f(14), f(−12)$.

2. The diagram shows part of a periodic function $f : x \to f(x)$, for $x \in \mathbf{R}$.

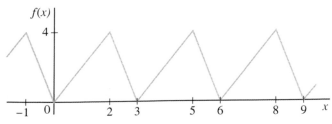

(i) Write down the period and range of $f(x)$.

(ii) Evaluate: $f(2), f(3), f(−1), f(−6)$.

3. The diagram shows part of a periodic function $f : x \to f(x)$, for $x \in \mathbf{R}$.

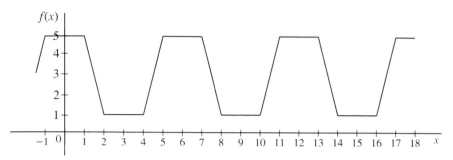

(i) Write down the period and range of $f(x)$.

(ii) Complete the following table:

x	1	2	3	8	11	-1	-6	-10	3.5	-6.5
$f(x)$										

4. The graph shows part of the graph of the periodic function $f : x \to f(x)$, for $x \in \mathbf{R}$.

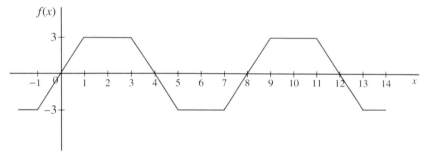

(i) Write down the period and range of $f(x)$.

(ii) Complete the following table:

x	1	4	6	8	20	-1	-8	-17	2.5	-1.5
$f(x)$										

5. The graph shows part of a periodic function $f : x \to f(x)$, for $x \in \mathbf{R}$.

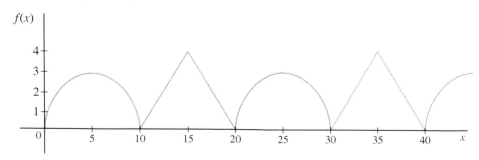

(i) Write down the period and range of $f(x)$.

(ii) Complete the following table:

x	5	10	15	20	45	60	−10	−25	−40	−50
$f(x)$										

(iii) Write down the three values of x in the domain $40 \leq x \leq 60$ for which $f(x) = 0$.

(iv) Write down the values of x in the domain $60 \leq x \leq 80$ for which:
  (a) $f(x)$ is a maximum   (b) $f(x)$ is increasing.

6. The diagram shows part of a periodic function $f : x \rightarrow f(x)$, for $x \in \mathbf{R}$.

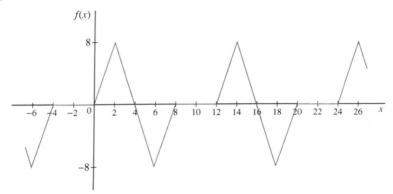

(i) Write down the period and range of the function.

(ii) Complete the following table:

x	0	2	4	6	8	10	12	30	41	−6	−10
$f(x)$											

(iii) Write down any four integer values of x in the domain $48 \leq x \leq 60$ for which $f(x) = 0$.

(iv) Write down the values of x in the domain $36 \leq x \leq 48$ for which:
  (a) $f(x)$ is a minimum   (b) $f(x)$ is decreasing   (c) $f(x)$ is positive and increasing.

7. The diagram shows part of the graph of the periodic function $f : x \rightarrow f(x)$ in the domain $0 \leq x \leq 6k$, $k \in \mathbf{R}$.

The period of $f(x)$ is 6 and its range is $[-2, 3]$.
Write down the values of k, a and b.

On separate diagrams draw sketches of $f(x)$ in the domain:
(i) $-3k \leq x \leq k$ (ii) $13k \leq x \leq 19k$.

Complete the following table:

x	1	4	6	7	−2	−10	−12	−16
$f(x)$								

If $f(k) = -2$, write down three values of k, $k \neq 1, 7$.

Simpson's Rule and Graphs

Simpson's Rule can be used to calculate the area between the graph of a function and the x-axis. Consider the graph of the function $f(x)$ in the domain $0 \leq x \leq 6$. Simpson's Rule can be used to calculate the area between the curve and the x-axis (shaded region). However, the negative y values must be taken as positive.

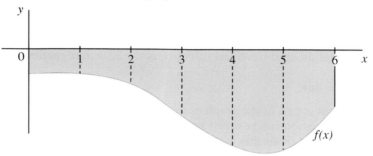

Exercise 7.7

1. The diagram shows the curve $f: x \rightarrow x^2 + 2$ in the domain $0 \leq x \leq 4$, $x \in \mathbf{R}$.

   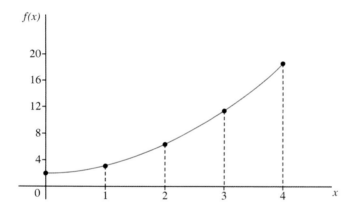

   (i) Complete the following table:

x	0	1	2	3	4
f(x)					

   (ii) Hence, use Simpson's Rule to calculate the area between the curve and the x-axis.

2. Graph the function $f: x \rightarrow 13 + 4x - x^2$, in the domain $0 \leq x \leq 6$, $x \in \mathbf{R}$.
   Use Simpson's Rule to calculate the area between the curve and the x-axis.

3. Graph the function $f: x \rightarrow 8 + 2x - x^2$, in the domain $-2 \leq x \leq 4$, $x \in \mathbf{R}$.
   Use Simpson's Rule to calculate the area between the curve and the x-axis.

4. Graph the function $f: x \rightarrow x^3 - 6x^2 + 9x + 6$, in the domain $0 \leq x \leq 4$, $x \in \mathbf{R}$.
   Use Simpson's Rule to calculate the area between the curve and the x-axis.

5. Graph the function $f: x \rightarrow x^3 + 3x^2 - x + 12$, in the domain $-4 \leq x \leq 2$, $x \in \mathbf{R}$.
   Use Simpson's Rule to calculate the area between the curve and the x-axis.

6. Graph the function $f: x \rightarrow -6 + 6x - x^2 - x^3$, in the domain $-3 \leq x \leq 3$, $x \in \mathbf{R}$.
   Use Simpson's Rule to calculate the area between the curve and the x-axis.

CHAPTER 8: DIFFERENTIATION

Differentiation from First Principles

Differentiation, or differential calculus, is the branch of mathematics measuring rates of change.

Slope of a line

On the right is part of the graph of the line $y = 3x$.
There is a relationship between x and y. For every increase in x, there is three times this increase in y.

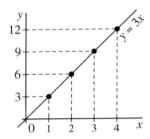

> Rate of change of y = 3 times rate of change of x

From coordinate geometry, the slope of the line $y = 3x$ is 3.

> Slope = 3

> Rate of change = Slope

The key word here is **slope**. The slope of a line will give **the rate of change** of the variable on the vertical axis with respect to the variable on the horizontal axis. Therefore, to find the rate of change we need only to find the slope.

Note: The y-axis is usually the vertical axis and the x-axis the horizontal axis. Therefore, the slope of a line will give the rate of change of y with respect to (the change in) x.

Slope of a curve

Consider the curve below and the tangents that are constructed on it.

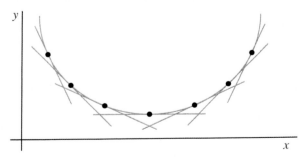

The slope of the curve at a point is equal to the slope of the tangent at that point. As we move along the curve the slope of each tangent changes. In other words, the rate of change of y with respect to x

changes. We need to find a method of finding the slope of the tangent at each point on the curve. The method of finding the slope of a tangent to a curve at any point on the curve is called **differentiation**.

Notation

We will now develop the method for finding the slope of the tangent to the curve $y = f(x)$ at any point $(x, f(x))$ on the curve.
Let the graph shown represent the function $y = f(x)$.
$(x, f(x))$ is a point on this curve and $(x + h, f(x + h))$ is a point further along the curve.
S is a line through these points.
T is a tangent to the curve at the point $(x, f(x))$.

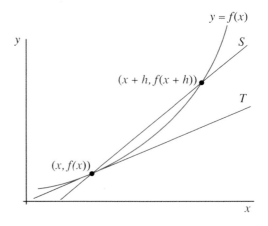

Slope of $S = \dfrac{y_2 - y_1}{x_2 - x_1}$

$= \dfrac{f(x+h) - f(x)}{(x+h) - h}$

$= \dfrac{f(x+h) - f(x)}{h}$

This would be a good approximation of the slope of the tangent T, at the point $(x, f(x))$, if $(x + h, f(x + h))$ is **very** close to $(x, f(x))$. By letting h get smaller, the point $(x + h, f(x + h))$ moves closer to $(x, f(x))$. The result is that the slope of S gets closer to the slope of T. In other words, as h approaches 0, the slope of S approaches the slope of T.

Mathematically speaking, we say that the slope of T is equal to the limit of the slope of S as h approaches 0. It is important to realise that h approaches 0, but **never** actually becomes equal to zero.

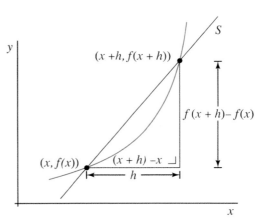

This is usually written $\lim\limits_{h \to 0} \dfrac{f(x+h) - f(x)}{h}$.

The process of finding this limiting value is called '**differentiation**'.

For neatness, this limit is written as $\dfrac{dy}{dx}$ (pronounced 'dee y, dee x') or $f'(x)$, pronounced 'f dash of x' or 'f prime of x'.

$\dfrac{dy}{dx}$ or $f'(x)$ is called the: '**differential coefficient**' or '**first derivative of y with respect to x**'.

The advantage of the notation $\dfrac{dy}{dx}$ is that it tells us which quantities are being compared.

$\dfrac{dy}{dx}$ is the derivative of y with respect to x.

$\dfrac{ds}{dt}$ is the derivative of s with respect to t.

$\dfrac{dA}{dr}$ is the derivative of A with respect to r.

Definition

> The slope of the tangent to the curve $y = f(x)$ at any point on the curve is given by:
> $$\frac{dy}{dx} = f'(x) = \lim_{h \to 0} \frac{f(x+h) - f(x)}{h}$$

Differentiation from first principles involves four steps:

Find:
1. $f(x+h)$
2. $f(x+h) - f(x)$
3. $\dfrac{f(x+h) - f(x)}{h}$
4. $\lim\limits_{h \to 0} \dfrac{f(x+h) - f(x)}{h}$

Example

Differentiate $x^2 - 3x + 2$ from first principles with respect to x.

Solution:

$$f(x) = x^2 - 3x + 2$$

1. $f(x+h) = (x+h)^2 - 3(x+h) + 2$ [replace x with $(x+h)$]
   $f(x+h) = x^2 + 2hx + h^2 - 3x - 3h + 2$ $[(x+h)^2 = x^2 + 2hx + h^2]$

2. $f(x+h) - f(x) = \cancel{x^2} + 2hx + h^2 - \cancel{3x} - 3h + \cancel{2} - \cancel{x^2} + \cancel{3x} - \cancel{2}$
   $f(x+h) - f(x) = 2hx + h^2 - 3h$

3. $\dfrac{f(x+h) - f(x)}{h} = 2x + h - 3$ [divide both sides by h]

4. $\lim\limits_{h \to 0} \dfrac{f(x+h) - f(x)}{h} = 2x + 0 - 3 = 2x - 3$ [take the limit]

Exercise 8.1

Differentiate each of the following with respect to x from first principles:

1. $x^2 + 2x + 3$
2. $x^2 + 3x - 2$
3. $x^2 - 2x + 5$
4. $x^2 - 5x - 3$
5. $x^2 - 4x - 4$
6. $x^2 + 5x$
7. $x^2 - 3x$
8. $x^2 + 6x$
9. $3x + 2$
10. $5x - 3$
11. $2x^2$
12. $2x^2 - 5x$
13. $3x^2 + 4$
14. $3x^2 - 2x$
15. $2x - x^2$
16. $3x - 2x^2$

Differentiation by Rule

Differentiation from first principles can become tedious and difficult. Fortunately, it is not always necessary to use first principles. There are a few rules (which can be derived from first principles) which enable us to write down the derivative of a function quite easily.

Rule 1: General Rule

$$y = x^n \text{ then } \frac{dy}{dx} = nx^{n-1}$$

$$y = ax^n \text{ then } \frac{dy}{dx} = nax^{n-1}$$

In words: Multiply by the power and reduce the power by 1.

Example

Differentiate each of the following with respect to x:

(i) $y = x^3$ (ii) $y = -5x^2$ (iii) $y = 4x$ (iv) $y = \frac{1}{x^5}$ (v) $y = 7$.

Solution:

(i) $y = x^3$ $\qquad \frac{dy}{dx} = 3x^{3-1} = 3x^2$

(ii) $y = -5x^2$ $\qquad \frac{dy}{dx} = 2 \times -5x^{2-1} = -10x^1 = -10x$

(iii) $y = 4x = 4x^1$ $\qquad \frac{dy}{dx} = 1 \times 4x^{1-1} = 4x^0 = 4 \quad (x^0 = 1)$

(iv) $y = \frac{1}{x^5} = x^{-5}$ $\qquad \frac{dy}{dx} = -5x^{-5-1} = -5x^{-6}$ or $-\frac{5}{x^6}$

(v) $y = 7 = 7x^0$ $\qquad \frac{dy}{dx} = 0 \times 7x^{0-1} = 0$

Part (v) leads to the rule:

The derivative of a constant = 0.

Note: The line $y = 7$ is a horizontal line. Its slope is 0. Therefore its derivative (also its slope) equals 0. In other words, the derivative of a constant always equals zero.

After practice, $\frac{dy}{dx}$ can be written down from inspection.

Sum or difference

If the expression to be differentiated contains more than one term just differentiate, separately, each term in the expression.

Example

(i) If $y = 3x^2 - 5x + 4$, find $\frac{dy}{dx}$.

(ii) If $f(x) = 4x^3 + x^2 - x - 6$, find $f'(x)$.

Solution:

(i) $y = 3x^2 - 5x + 4$
$\frac{dy}{dx} = 6x - 5$

(ii) $f(x) = 4x^3 + x^2 - x - 6$
$f'(x) = 12x^2 + 2x - 1$

Exercise 8.2

Find $\frac{dy}{dx}$ if:

1. $y = x^4$
2. $y = x^6$
3. $y = 3x^2$
4. $y = -5x^4$
5. $y = 4x$
6. $y = -3x$
7. $y = 8$
8. $y = -5$
9. $y = \frac{1}{x^3}$
10. $y = \frac{1}{x}$
11. $y = x^4 + 2x^3$
12. $y = 2x^3 + 5x^2$
13. $y = 3x^2 + 4x$
14. $y = 2x^2 - 6x$
15. $y = 5x - 2x^2$
16. $y = 3x - 1$
17. $y = x^3 + 2x^2 + 5x$
18. $y = x - 3x^2 - 4x^3$
19. $y = 4 - 5x^2 - 6x^4$

Find $f'(x)$ if:

20. $f(x) = x^2 - x - 6$
21. $f(x) = x^3 - 3x^2 + 4$
22. $f(x) = 20x - 2x^2$
23. $f(x) = 2x^3 - 8x^2 + 7x - 6$
24. $f(x) = x^3 - 2x^2 + 4x - 1$
25. $f(x) = 8 + 2x - 3x^2 - x^3$
26. $f(x) = x^3 + \frac{1}{x^3}$
27. $3x^2 + \frac{1}{x^2}$
28. $f(x) = x^4 + 4 + \frac{1}{x^4}$

Evaluating Derivatives

Often we may be asked to find the value of the derivative for a particular value of the function.

DIFFERENTIATION

Example

(i) If $y = 2x^3 - 4x + 3$, find the value of $\dfrac{dy}{dx}$ when $x = 1$.

(ii) If $s = 4t^2 + 10t - 7$, find the value of $\dfrac{ds}{dt}$ when $t = -2$.

Solution:

(i) $y = 2x^3 - 4x + 3$

$\dfrac{dy}{dx} = 6x^2 - 4$

$\left.\dfrac{dy}{dx}\right|_{x=1} = 6(1)^2 - 4$

$= 6 - 4 = 2$

(ii) $s = 4t^2 + 10t - 7$

$\dfrac{ds}{dt} = 8t + 10$

$\left.\dfrac{ds}{dt}\right|_{t=-2} = 8(-2) + 10$

$= -16 + 10 = -6$

Exercise 8.3

1. If $y = 3x^2 + 4x + 2$, find the value of $\dfrac{dy}{dx}$ when $x = 1$.

2. If $y = 2x^3 + 4x^2 + 3x - 5$, find the value of $\dfrac{dy}{dx}$ when $x = 2$.

3. If $y = 4x^3 - 3x^2 + 5x - 3$, find the value of $\dfrac{dy}{dx}$ when $x = -1$.

4. If $s = 3t - 2t^2$, find $\dfrac{ds}{dt}$ when $t = 2$.

5. If $s = t^3 - 2t^2 - t + 1$, find $\dfrac{ds}{dt}$ when $t = -1$.

6. If $A = 3r^2 - 5r$, find $\dfrac{dA}{dr}$ when $r = 3$.

7. If $V = 3h - h^2 - 3h^3$, find $\dfrac{dV}{dh}$ when $h = 2$.

8. If $h = 20t - 5t^2$, find $\dfrac{dh}{dt}$ when $t = 4$.

9. If $A = \pi r^2$, find $\dfrac{dA}{dr}$ when $r = 5$, leaving your answer in terms of π.

10. If $V = \tfrac{4}{3}\pi r^3$, find $\dfrac{dV}{dr}$ when $r = 3$, leaving your answer in terms of π.

Rule 2: Product Rule

Suppose u and v are functions of x.

If $y = uv$,

then $\dfrac{dy}{dx} = u\dfrac{dv}{dx} + v\dfrac{du}{dx}$

In words: First by the derivative of the second + second by the derivative of the first.

Note: The word '**product**' refers to quantities being multiplied.

Example

If $y = (x^2 - 3x + 2)(x^2 - 2)$, find $\dfrac{dy}{dx}$. Hence, evaluate $\dfrac{dy}{dx}$ when $x = -1$.

Solution:

Let $u = x^2 - 3x + 2$ and let $v = x^2 - 2$.

$\dfrac{du}{dx} = 2x - 3$ and $\dfrac{dv}{dx} = 2x$

$\dfrac{dy}{dx} = u\dfrac{dv}{dx} + v\dfrac{du}{dx}$ (product rule)

$= (x^2 - 3x + 2)(2x) + (x^2 - 2)(2x - 3)$

$= 2x^3 - 6x^2 + 4x + 2x^3 - 3x^2 - 4x + 6$

$= 4x^3 - 9x^2 + 6$

$\left.\dfrac{dy}{dx}\right|_{x=-1} = 4(-1)^3 - 9(-1)^2 + 6 = -4 - 9 + 6 = -7$

Exercise 8.4

Use the product rule to find $\dfrac{dy}{dx}$ if:

1. $y = (2x + 4)(3x + 5)$
2. $y = (x^2 + 2)(2x^2 + 7)$
3. $y = (2x + 3)(x^2 + 3x + 4)$
4. $y = (x^2 + 3x)(x^3 - 3x + 4)$
5. $y = (3x^2 - 5x + 3)(3x - 2)$
6. $y = (x + 3)(x - 2)$
7. $y = (x^3 - 2x)(3 - 2x)$
8. $y = (x - 4)(x^3 - 5)$
9. $y = (x^2 + x + 1)(x - 2)$
10. $y = (2x - x^3 - x^4)(x^2 - 3)$
11. $y = (x^3 - x^2 - x)(x^3 + 2x)$
12. $y = (5x^3 - 6x)(x^2 - 3x - 1)$

13. Let $f(x) = (x^2 - 2x)(3x + 2)$. Find $f'(x)$, the derivative of $f(x)$.

14. Let $f(x) = (x^3 - 1)(2x - x^2)$. Find $f'(x)$, the derivative of $f(x)$.

15. If $y = (2x - x^2)(x^2 - x - 1)$, evaluate $\dfrac{dy}{dx}$ when $x = 0$.

16. If $s = (3t^2 - 4t)(t^2 - 4)$, find $\dfrac{ds}{dt}$ and evaluate it when $t = 1$.

17. If $x = (2h^2 - 3h + 5)(h - 2)$, evaluate $\dfrac{dx}{dh}$ when $h = 2$.

18. Find the coefficient of x^3 in the derivative of $(2x^2 - x - 3)(1 - 2x^2)$ with respect to x.

Rule 3: Quotient Rule

Suppose u and v are functions of x.

If $y = \dfrac{u}{v}$,

then $\dfrac{dy}{dx} = \dfrac{v\dfrac{du}{dx} - u\dfrac{dv}{dx}}{v^2}$.

In words: $\dfrac{\text{Bottom by the derivative of the top} - \text{Top by the derivative of the bottom}}{(\text{Bottom})^2}$

Note: Quotient is another name for a fraction. The quotient rule refers to one quantity divided by another.

Example

If $y = \dfrac{x^2}{x + 2}$ find $\dfrac{dy}{dx}$ and, hence, find the value of $\dfrac{dy}{dx}$ when $x = 2$.

Solution:

$$y = \dfrac{x^2}{x + 2}$$

Let $u = x^2$ and $v = x + 2$

$\dfrac{du}{dx} = 2x$ and $\dfrac{dv}{dx} = 1$

$$\frac{dy}{dx} = \frac{v\dfrac{du}{dx} - u\dfrac{dv}{dx}}{v^2} \quad \text{(quotient rule)}$$

$$= \frac{(x+2)(2x) - (x^2)(1)}{(x+2)^2}$$

$$= \frac{2x^2 + 4x - x^2}{(x+2)^2}$$

$$= \frac{x^2 + 4x}{(x+2)^2}$$

Note: It is usual practice to simplify the top but **not** the bottom.

$$\left.\frac{dy}{dx}\right|_{x=2} = \frac{(2)^2 + 4(2)}{(2+2)^2} = \frac{4+8}{(4)^2} = \frac{12}{16} = \frac{3}{4}$$

Exercise 8.5

Find $\dfrac{dy}{dx}$ if:

1. $y = \dfrac{3x+2}{x+1}$
2. $y = \dfrac{2x+1}{x+3}$
3. $y = \dfrac{x}{x-1}$
4. $y = \dfrac{5x+2}{x+4}$
5. $y = \dfrac{1}{x+2}$
6. $y = \dfrac{1}{x-3}$
7. $y = \dfrac{3}{x+4}$
8. $y = \dfrac{x^2}{x-1}$
9. $y = \dfrac{2x+3}{5x-4}$
10. $y = \dfrac{2x-3}{x^2+1}$
11. $y = \dfrac{2x^2}{3x^2-1}$
12. $y = \dfrac{x^2+2}{4-x^2}$

13. If $y = \dfrac{3x+2}{x-2}$, evaluate $\dfrac{dy}{dx}$ at $x = 1$.

14. If $y = \dfrac{2x^2-1}{x+5}$, evaluate $\dfrac{dy}{dx}$ at $x = 2$.

15. If $y = \dfrac{x^2-4x}{5x-1}$, evaluate $\dfrac{dy}{dx}$ at $x = 0$.

Rule 4: Chain Rule

The chain rule is used when the given function is raised to a power, e.g. $y = (x^2 - 3x + 4)^4$.

To differentiate using the chain rule, do the following in *one* step:

> (a) Treat what is inside the bracket as a single variable and differentiate this (multiply by the power and reduce the power by one).
>
> (b) Multiply this result by the derivative of what is inside the bracket.

$$\text{If } y = (\text{function})^n,$$
$$\text{then } \frac{dy}{dx} = n \, (\text{function})^{n-1} \, (\text{derivative of the function}).$$

Example

Find $\frac{dy}{dx}$ if: **(i)** $y = (x^2 + 3x)^5$ **(ii)** $y = (2x^2 - 5x + 3)^{20}$.

Solution:

(i) $y = (x^2 + 3x)^5$

$\frac{dy}{dx} = 5(x^2 + 3x)^4(2x + 3)$

(ii) $y = (2x^2 - 5x + 3)^{20}$

$\frac{dy}{dx} = 20(2x^2 - 5x + 3)^{19}(4x - 5)$

Exercise 8.6

Find $\frac{dy}{dx}$ if:

1. $y = (2x + 3)^5$
2. $y = (5x - 1)^4$
3. $y = (x^2 + 3x)^3$
4. $y = (x^2 - 5x - 6)^7$
5. $y = (4 - 5x)^6$
6. $y = (3 - 2x)^5$
7. $y = (1 + x^2)^4$
8. $y = (5 - 2x^2)^7$
9. $y = (4 - 3x - x^2)^8$
10. $y = \left(x^2 + \frac{1}{x^2}\right)^5$
11. $y = \left(x^3 - \frac{1}{x^3}\right)^4$
12. $y = \left(1 + \frac{1}{x}\right)^{10}$

13. If $y = (x^2 - 1)^4$, evaluate $\frac{dy}{dx}$ when $x = 1$.

14. If $y = (2x^2 - 3x + 1)^{10}$, find the value of $\frac{dy}{dx}$ when $x = 0$.

15. If $y = (h^2 - h + 1)^2$, find the value of $\frac{dy}{dh}$ when $h = 1$.

16. If $y = (x^2 + 1)^3$, find the value of $\frac{dy}{dx}$ when $x = 1$.

17. If $y = (2t^2 + 3t - 1)^8$, find the value of $\dfrac{dy}{dt}$ when $t = -2$.

18. If $f(x) = (2x - 3)^4$, evaluate $f'(x)$ at $x = \tfrac{3}{2}$.

Finding the Slope and Equation of a Tangent to a Curve at a Point on the Curve

$\dfrac{dy}{dx}$ = the slope of a tangent to a curve at any point on the curve.

To find the slope and equation of a tangent to a curve at a given point (x_1, y_1), on the curve, do the following:

Step 1: Find $\dfrac{dy}{dx}$.

Step 2: Evaluate $\dfrac{dy}{dx}\bigg|_{x=x_1}$ (this gives the slope of the tangent, m)

Step 3: Use m (from step 2) and the given point (x_1, y_1) in the equation:
$$(y - y_1) = m(x - x_1)$$

Note: Sometimes only the value of x is given. When this happens, substitute the value of x into the original function to find y for step 3.

Example ▼

Find the equation of the tangent to the curve $y = 3 + 2x - x^2$ at the point $(2, 3)$.

Solution:

$$y = 3 + 2x - x^2$$

Step 1: $\dfrac{dy}{dx} = 2 - 2x$

Step 2: At the point $(2, 3)$, $x = 2$

$$m = \dfrac{dy}{dx}\bigg|_{x=2} = 2 - 2(2) = 2 - 4 = -2$$

Step 3: $m = -2,\quad x_1 = 2,\quad y_1 = 3$

$$(y - y_1) = m(x - x_1)$$
$$(y - 3) = -2(x - 2)$$
$$y - 3 = -2x + 4$$
$$2x + y - 3 - 4 = 0$$
$$2x + y - 7 = 0$$

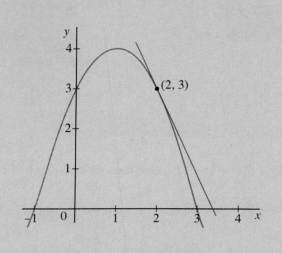

Exercise 8.7

1. Find the slope of the tangent to the curve $x^2 - 3x + 2$ at the point $(1, 0)$.
2. Find the slope of the tangent to the curve $y = 3 - 3x - x^2$ at the point $(2, -7)$.
3. Find the slope of the tangent to the curve $y = x^2 - 2x - 3$ at the point $(1, -4)$.
4. Find the equation of the tangent to the curve $y = x^2 - x - 2$ at the point $(2, 0)$.
5. Find the equation of the tangent to the curve $y = x^3 - 2x^2 - 4x + 1$ at the point $(-1, 2)$.
6. Find the equation of the tangent to the curve $y = (2x + 3)^3$ at $x = -1$.
7. Find the equation of the tangent to the curve $y = \dfrac{2x - 4}{x - 4}$ at $x = 6$.
8. Verify that the point $(-1, -2)$ is on the curve $y = \dfrac{x - 1}{x + 2}$ and find the equation of the tangent to the curve at this point.
9. Show that the points $(2, 0)$ and $(3, 0)$ are on the curve $y = x^2 - 5x + 6$.
   Find the equations of the tangents to the curve at these points and investigate if these two tangents are at right angles to each other.
10. Show that the tangent to the curve $y = 3x^2 - 4x + 11$ at the point $(1, 10)$ is parallel to the line $2x - y + 5 = 0$.
11. Show that the tangent to the curve $y = \dfrac{6x - 3}{4x + 2}$ at the point $(1, \tfrac{1}{2})$ is parallel to the line $2x - 3y - 6 = 0$.

Given $\dfrac{dy}{dx}$, to Find the Coordinates of the Corresponding Points on a Curve

Sometimes the value of $\dfrac{dy}{dx}$ (slope of the curve at any point on it) is given and we need to find the coordinates of the point, or points, corresponding to this slope.

When this happens do the following:

Step 1: Find $\dfrac{dy}{dx}$.

Step 2: Let $\dfrac{dy}{dx}$ equal the given value of the slope and solve this equation for x.

Step 3: Substitute the x values obtained in step 2 into the original function to get the corresponding values of y.

Example

Find the coordinates of the points on the curve $y = x^3 - 3x^2 - 8x + 5$ at which the tangents to the curve make angles of 45° with the positive sense of the x-axis.

Solution:
$$y = x^3 - 3x^2 - 8x + 5$$

Step 1: $\dfrac{dy}{dx} = 3x^2 - 6x - 8$

Step 2: Angle of 45° $\Rightarrow$ slope = 1   (as tan 45° = 1)

Let $\dfrac{dy}{dx} = 1$ (given slope in disguise)

$$3x^2 - 6x - 8 = 1$$
$$3x^2 - 6x - 9 = 0$$
$$x^2 - 2x - 3 = 0$$
$$(x - 3)(x + 1) = 0$$
$$x - 3 = 0 \quad \text{or} \quad x + 1 = 0$$
$$x = 3 \quad \text{or} \quad x = -1$$

Step 3: Find the y values.

$y = x^3 - 3x^2 - 8x + 5$	$y = x^3 - 3x^2 - 8x + 5$
$x = 3$	$x = -1$
$y = (3)^3 - 3(3)^2 - 8(3) + 5$	$y = (-1)^3 - 3(-1)^2 - 8(-1) + 5$
$= 27 - 27 - 24 + 5$	$= -1 - 3 + 8 + 5$
$= -19$	$= 9$
point $(3, -19)$	point $(-1, 9)$

Thus the required points are $(3, -19)$ and $(-1, 9)$.

Note: $\tan 135° = -1$.

Example

Find the coordinates of the point on the curve $y = x^2 - x$ where the tangent to the curve is parallel to the line $y = 3x - 5$.

Step 1: Curve: $y = x^2 - x$ Line: $y = 3x - 5$

$\dfrac{dy}{dx} = 2x - 1$ $\dfrac{dy}{dx} = 3$

Step 2: Slope of curve = Slope of line (given)

$$2x - 1 = 3$$
$$2x = 4$$
$$x = 2$$

Step 3: $y = x^2 - x$

$x = 2$

$y = (2)^2 - 2$

$= 4 - 2 = 2$

Thus, the required point is $(2, 2)$

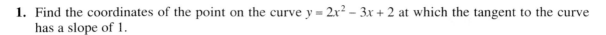

Exercise 8.8

1. Find the coordinates of the point on the curve $y = 2x^2 - 3x + 2$ at which the tangent to the curve has a slope of 1.

2. Find the coordinates of the point on the curve $y = 3x^2 - 2x + 5$ at which the tangent to the curve has a slope of -10.

3. Find the coordinates of the point on the curve $y = x^2 - 5x + 3$ at which the tangent to the curve has a slope of 3.

4. Find the coordinates of the points on the curve $y = x^3 - 3x^2$ at which the tangents to the curve have a slope of 9.

5. Find the coordinates of the points on the curve $y = 2x^3 - 3x^2 - 13x + 2$ at which the tangents to the curve make angles of $135°$ with the positive sense of the x-axis.

6. Find the coordinates of the point on the curve $y = 2x^2 - 2x + 5$ at which the tangents to the curve are parallel to the line $y = 2x - 3$.

7. Find the coordinates of the points on the curve $y = 2x^3 - 3x^2 - 12x$ at which the tangents to the curve are parallel to the line $y = 24x + 3$.

8. Find the coordinates of the points on the curve $y = \dfrac{x}{1+x}$ at which the tangents to the curve are parallel to the line $x - y + 8 = 0$.
   Find the equations of the two tangents at these points.

9. (i) What is the slope of the x-axis?
   (ii) Show that the tangent of the curve $f(x) = x^2 - 2x + 5$, at the point where $x = 1$, is parallel to the x-axis.
   (iii) For what value of x is the tangent to the graph of $y = x^2 - 6x + 5$ parallel to the x-axis?
   (iv) Find the values of x for which the tangents to the graph of $y = \dfrac{x^2 + 2}{2x - 1}$ are parallel to the x-axis.

10. Find the value of a, if the slope of the tangent to the curve $y = x^2 + ax$ is 2 at the point where $x = 3$.

11. Let $f(x) = x^3 + ax^2 + 2$ for all $x \in \mathbf{R}$ and for $a \in \mathbf{R}$.
    The slope of the tangent to the curve $y = f(x)$ at $x = -1$ is 9. Find the value of a.

12. Let $g(x) = (x + 1)(x^2 - 4x)$ for $x \in \mathbf{R}$.
    (i) For what two values of x is the slope of the tangent to the curve of $g(x)$ equal to 20?
    (ii) Find the equations of the two tangents to the curve of $g(x)$ which have slope 20.

13. Let $f(x) = 2x^3 - 9x^2 + 12x + 1$ for $x \in \mathbf{R}$.
    Find $f'(x)$, the derivative of $f(x)$. At the two points, (x_1, y_1) and (x_2, y_2), the tangents to the curve $y = f(x)$ are parallel to the x-axis, where $x_2 > x_1$.
    Show that: (i) $x_1 + x_2 = 3$ (ii) $y_1 = y_2 + 1$.

Maximum and Minimum Points

$\dfrac{dy}{dx}$ can also be used to find the local maximum or local minimum points on a curve.

On the right is part of the graph of a cubic function. At the turning points, a and b, the tangents to the curve are horizontal (parallel to the x-axis). In other words, at these points the slope of the tangent is zero. These turning points are also called the **local maximum point**, point a, and the **local minimum point**, point b.

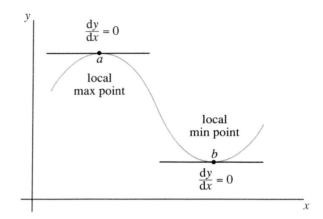

At a maximum or minimum point, $\dfrac{dy}{dx} = 0$.

To find the maximum or minimum points on a curve, do the following:

Step 1: Find $\dfrac{dy}{dx}$.

Step 2: Let $\dfrac{dy}{dx} = 0$ and solve this equation for x.

Step 3: Substitute x values obtained in step 2 into the original function to get the corresponding values of y.

Step 4: By comparing the y values we can determine which point is the local maximum or minimum point. The point with the greater y value is the local maximum point and vice versa.

Note: The graph of a quadratic function $f(x) = ax^2 + bx + c$ has only one turning point.

1. If $a > 0$, the turning point will be a minimum.

2. If $a < 0$, the turning point will be a maximum.

Example

Find, using calculus, the coordinates of the local maximum point and the local minimum point of the curve $y = x^3 - 6x^2 + 9x + 4$.

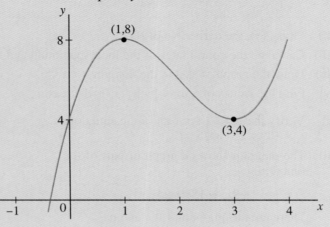

Graph of $y = x^3 - 6x^2 + 9x + 4$.

Solution:

Step 1: $y = x^3 - 6x^2 + 9x + 4$

$\dfrac{dy}{dx} = 3x^2 - 12x + 9$

Step 2: Let $\dfrac{dy}{dx} = 0$.

$3x^2 - 12x + 9 = 0$

$x^2 - 4x + 3 = 0$

$(x - 3)(x - 1) = 0$

$x - 3 = 0$ or $x - 1 = 0$

$x = 3$ or $x = 1$

Step 3: Find the y values.

$y = x^3 - 6x^2 + 9x + 4$

$x = 3$

$y = (3)^3 - 6(3)^2 + 9(3) + 4$

$= 27 - 54 + 27 + 4$

$= 4$

point $(3, 4)$

$y = x^3 - 6x^2 + 9x + 4$

$x = 1$

$y = (1)^3 - 6(1)^2 + 9(1) + 4$

$= 1 - 6 + 9 + 4$

$= 8$

point $(1, 8)$

Step 4:   8 is greater than 4.

Thus, the local maximum point is $(1, 8)$ and the local minimum point is $(3, 4)$.

Exercise 8.9

Find, using calculus, the coordinates of the local minimum point of each of the following curves:

1. $y = x^2 - 4x + 3$
2. $y = x^2 + 6x + 1$
3. $y = 2x^2 - 8x + 3$

Find, using calculus, the coordinates of the local maximum point of each of the following curves:

4. $y = 7 - 6x - x^2$
5. $y = 5 - 4x - x^2$
6. $y = 1 - 12x - 3x^2$

Find, using calculus, the coordinates of the local maximum and the local minimum points of each of the following curves (in each case distinguish between the maximum and the minimum):

7. $y = x^3 - 3x^2 - 9x + 4$
8. $y = x^3 - 6x^2 + 9x - 1$
9. $y = 2x^3 - 9x^2 + 12x + 1$
10. $y = 8 + 9x - 3x^2 - x^3$
11. $y = 5 - 12x - 9x^2 - 2x^3$
12. $y = x^3 - 12x + 4$

13. Let $f(x) = x^3 - 9x^2 + 24x - 17$ for $x \in \mathbf{R}$.
    (i) Complete the following table:

x	0	1	2	3	4	5	6
$f(x)$		-1				3	

    (ii) Find $f'(x)$, the derivative of $f(x)$.
    (iii) Calculate the coordinates of the local maximum and the local minimum points of $f(x)$.
    (iv) Draw the graph of $f(x)$ in the domain $0 \leqslant x \leqslant 6$.
    (v) Find the values of x for which $f(x)$ is decreasing.

14. (i) Verify that $x = -1$ is a root of the equation:
    $x^3 - 9x^2 + 15x + 25 = 0$.
    (ii) The diagram shows a graph of part of the function:

    $f : x \rightarrow x^3 - 9x^2 + 15x + 25$

    There are turning points at c and d.
    The curve intersects the x-axis at a and d and the y-axis at b.
    Find the coordinates of a, b, c and d.

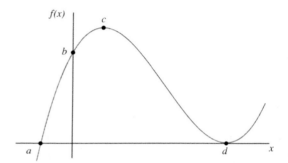

15. Let $f(x) = x^3 + ax^2 + 2$, for all $x \in \mathbf{R}$ and for $a \in \mathbf{R}$.
    $f(x)$ has a turning point (a local maximum or a local minimum) at $x = 2$.
    (i) Find the value of a and the coordinates of the turning point at $x = 2$.
    (ii) Find the coordinates of the other turning point of $f(x)$.
    (iii) Draw the graph of $f(x)$ in the domain $-2 \leqslant x \leqslant 4$.
    (iv) Find the range of values of k for which $f(x) = k$ has three roots.

16. Let $f(x) = px^2 + qx + r$, for all $x \in \mathbf{R}$ and $p, q, r \in \mathbf{R}$.
    Find $f'(x)$, the derivative of $f(x)$, in terms of p, q and x.

17. Let $f(x) = ax^3 + bx + c$, for all $x \in \mathbf{R}$ and for $a, b, c \in \mathbf{R}$.
    Use the information which follows to find the value of a, of b and of c:
    (i) $f(0) = -3$
    (ii) the slope of the tangent to the curve of $f(x)$ at $x = 2$ is 18
    (iii) the curve of $f(x)$ has a local minimum point at $x = 1$.

Increasing and Decreasing

$\dfrac{dy}{dx}$, being the slope of a tangent to a curve at any point on the curve, can be used to determine if, and where, a curve is increasing or decreasing.

Note: Graphs are read from left to right.

Where a curve is increasing, the tangent to the curve will have a positive slope.

Therefore, where a curve is increasing, $\dfrac{dy}{dx}$ will be positive.

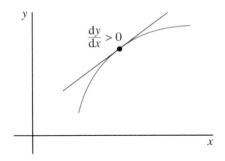

Where a curve is decreasing, the tangent to the curve will have a negative slope.

Therefore, where a curve is decreasing, $\dfrac{dy}{dx}$ will be negative.

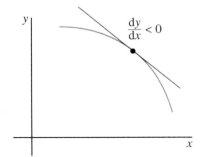

Example

If $y = \dfrac{2x}{1-x}$, show that $\dfrac{dy}{dx} > 0$ for all $x \neq 1$.

Solution:

$$y = \dfrac{2x}{1-x}$$

$$\dfrac{dy}{dx} = \dfrac{(1-x)(2) - (2x)(-1)}{(1-x)^2} \quad \text{(quotient rule)}$$

$$= \dfrac{2 - 2x + 2x}{(1-x)^2}$$

$$= \dfrac{2}{(1-x)^2}$$

$$\therefore \dfrac{dy}{dx} = \dfrac{2}{(1-x)^2}$$

$(1-x)^2 > 0$ for all $x \neq 1$, $\quad 2 > 0 \quad$ (top and bottom both positive)

$\therefore \dfrac{2}{(1-x)^2} > 0$ for all $x \neq 1$.

$\therefore \dfrac{dy}{dx} > 0$ for all $x \neq 1$.

Note: (any real number)^2 will always be a positive number unless the number is zero.

$\therefore (1-x)^2$ must always be positive, unless $x = 1$ which gives $0^2 = 0$.

Exercise 8.10

1. Let $f(x) = x^2 - 2x - 8$. Find the values of x for which $f(x)$ is: **(i)** decreasing **(ii)** increasing.

2. Let $f(x) = 12 - 6x - x^2$. Find the values of x for which $f(x)$ is: **(i)** increasing **(ii)** decreasing.

3. Let $f(x) = x^3 + 4x - 2$. Show that $\dfrac{dy}{dx} > 0$ for all $x \in \mathbf{R}$.

4. Let $f(x) = 5 - 2x - x^3$. Show that $\dfrac{dy}{dx} < 0$ for all $x \in \mathbf{R}$.

5. Show that the curve $y = x^2 - 3x + 5$ is increasing at the point $(2, 3)$.

6. Show that the curve $y = 10 - x - 2x^2$ is decreasing at the point $(3, -11)$.

7. If $y = \dfrac{x}{1 - x^2}$, show that $\dfrac{dy}{dx} > 0$ for all $x \neq \pm 1$.

8. If $y = \dfrac{x + 2}{x - 1}$, show that $\dfrac{dy}{dx} < 0$ for all $x \neq 1$.

9. If $y = \dfrac{4x + 1}{x - 3}$, show that $\dfrac{dy}{dx} < 0$ for all $x \neq 3$.

10. If $y = \dfrac{2x - 5}{x + 2}$, show that $\dfrac{dy}{dx} > 0$ for all $x \neq -2$.

11. An artificial ski-slope is described by the function $h = 2 - 8s - 4s^2 - \tfrac{2}{3}s^3$, where s is the horizontal distance and h is the height of the slope. Show that the slope is all downhill.

Rates of Change

The derivative $\dfrac{dy}{dx}$ is called the 'rate of change of y with respect to x'.

It shows how changes in y are related to changes in x.

If $\dfrac{dy}{dx} = 4$, then y is increasing 4 times as fast as x increases.

If $\dfrac{dy}{dx} = -5$, then y is decreasing 5 times as fast as x increases.

The derivative $\dfrac{dh}{dt}$ is called the 'rate of change of h with respect to t'.

The derivative $\dfrac{dR}{dV}$ is called the 'rate of change of R with respect to V'.

DIFFERENTIATION

If s denotes the displacement (position) of a particle, from a fixed point p at time t, then:

> 1. Velocity $= v = \dfrac{ds}{dt}$,
> the rate of change of position with respect to time.
>
> 2. Acceleration $= a = \dfrac{dv}{dt} = \dfrac{d^2s}{dt^2}$,
> the rate of change of velocity with respect to time.

To find $\dfrac{d^2s}{dt^2}$, simply find $\dfrac{ds}{dt}$ and differentiate this.
In other words, differentiate twice.

Note: 'Speed' is often used instead of 'velocity'. However, speed can never be negative, whereas velocity can be negative.

If $\dfrac{ds}{dt} > 0$, the particle is moving away from p (distance from p is increasing).

If $\dfrac{ds}{dt} < 0$, the particle is moving towards p (distance from p is decreasing).

Example ▼

A particle moves along a straight line such that, after t seconds, the distance, s metres, is given by $s = t^3 - 9t^2 + 15t - 3$. Find:

(i) the velocity and acceleration of the particle in terms of t

(ii) the values of t when its velocity is zero

(iii) the acceleration after $3\tfrac{1}{2}$ seconds

(iv) the time at which the acceleration is 6 m/s² and the velocity at this time.

Solution:

(i) $\qquad s = t^3 - 9t^2 + 15t - 3$

$\qquad\qquad v = \dfrac{ds}{dt} = 3t^2 - 18t + 15 \qquad$ (velocity at any time t)

$\qquad\qquad a = \dfrac{d^2s}{dt^2} = 6t - 18 \qquad$ (acceleration at any time t)

(ii) Values of t when velocity is zero

$$\text{velocity} = 0$$

$$\therefore \quad \frac{ds}{dt} = 0$$

$$\therefore \quad 3t^2 - 18t + 15 = 0$$

$$t^2 - 6t + 5 = 0$$

$$(t-1)(t-5) = 0$$

$$t - 1 = 0 \quad \text{or} \quad t - 5 = 0$$

$$t = 1 \quad \text{or} \quad t = 5$$

Thus, the particle is stopped after 1 second and again after 5 seconds.

(iii) Acceleration after $3\frac{1}{2}$ seconds

$$\text{acceleration} = \frac{d^2s}{dt^2} = 6t - 18,$$

When $t = 3\frac{1}{2}$,

$$\text{acceleration} = 6(3\frac{1}{2}) - 18$$

(put in $t = 3\frac{1}{2}$)

$$= 21 - 18$$

$$= 3$$

$\therefore$ Acceleration after $3\frac{1}{2}$ seconds = 3 m/s^2.

(iv) Time at which acceleration is 6 m/s^2

$$\text{acceleration} = 6 \text{ m/s}^2$$

$$\therefore \quad \frac{d^2s}{dt^2} = 6$$

$$\therefore \quad 6t - 18 = 6$$

$$6t = 24$$

$$t = 4$$

After 4 seconds the acceleration is 6 m/s^2.

Velocity after 4 seconds

$$\text{velocity} = 3t^2 - 18t + 15$$

when $t = 4$,

$$\text{velocity} = 3(4)^2 - 18(4) + 15$$

(put in $t = 4$)

$$= 48 - 72 + 15$$

$$= -9 \text{ m/s}$$

After 4 seconds the velocity is -9 m/s.

The negative value means it is going in the opposite direction to which it started, after 4 seconds.

Example ▼

A ball is thrown vertically up in the air. The height, h metres, reached above the ground t seconds after it was thrown is given by $h = 8t - t^2$.

Find:

(i) the height of the ball after 3 seconds

(ii) the speed of the ball after 1 second

(iii) the height of the ball when its speed is 4 m/s.

(iv) After how many seconds does the ball just begin to fall back downwards?

Solution:

(i) Height at $t = 3$

$$h = 8t - t^2$$

When $t = 3$,

$$h = 8(3) - (3)^2$$

(put in $t = 3$)

$$h = 24 - 9$$
$$h = 15$$

∴ Height after 3 seconds = 15 m.

(ii) Speed after 1 second

$$h = 8t - t^2$$
$$\frac{dh}{dt} = 8 - 2t$$

(this is the speed in terms of t)

∴ speed after 1 second

$$= 8 - 2(1) = 8 - 2 = 6$$

(put in $t = 1$)

∴ Speed after 1 second = 6 m/s.

(iii) Height of the ball when the speed is 4 m/s

Given: speed = 4

∴ $\dfrac{dh}{dt} = 4$

∴ $8 - 2t = 4$

$$-2t = -4$$
$$2t = 4$$
$$t = 2$$

$$h = 8t - t^2$$

When $t = 2$,

$$h = 8(2) - (2)^2$$

(put in $t = 2$)

$$h = 16 - 4 = 12 \text{ m}$$

∴ When the speed is 4 m/s, the height of the ball = 12 m.

(iv) After how many seconds does the ball begin to fall downwards?

The ball begins to fall downwards when it has reached its maximum height. It reaches its maximum height when its speed = 0.

speed = 0

∴ $\dfrac{dh}{dt} = 0$

∴ $8 - 2t = 0$

$$-2t = -8$$
$$2t = 8$$
$$t = 4$$

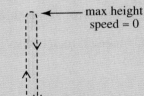

max height
speed = 0

∴ Ball begins to fall downwards after 4 seconds.

Exercise 8.11

1. If $s = t^3 - 2t^2$, evaluate: (i) $\dfrac{ds}{dt}$ at $t = 3$ (ii) $\dfrac{d^2s}{dt^2}$ at $t = 2$.

2. If $h = 4t^3 - 12t + 8$, evaluate: (i) $\dfrac{dh}{dt}$ at $t = 1$ (ii) $\dfrac{d^2h}{dt^2}$ at $t = \tfrac{1}{2}$.

3. A ball bearing rolls along the ground. It starts to move at $t = 0$ seconds. The distance that it has travelled at t seconds is given by $s = t^3 - 6t^2 + 9t$.
   Find:
   (i) $\dfrac{ds}{dt}$ and $\dfrac{d^2s}{dt^2}$, its speed and acceleration, in terms of t
   (ii) the speed of the ball bearing when $t = 4$ seconds
   (iii) the acceleration of the ball bearing when $t = 3$ seconds
   (iv) the times at which the speed is zero
   (v) the time at which the acceleration is zero
   (vi) the time at which the acceleration is 6 m/s^2
   (vii) the time at which the speed is 24 m/s.

4. A particle moves along a straight line such that, after t seconds, the distance s metres from a fixed point o is given by $s(t) = t^3 - 9t^2 + 24t$, $t \geq 0$.
   Find:
   (i) $\dfrac{ds}{dt}$ and $\dfrac{d^2s}{dt^2}$, its speed and acceleration, in terms of t
   (ii) the speed of the particle after 6 seconds
   (iii) the times when the speed is zero
   (iv) the acceleration of the particle after 4 seconds
   (v) the time at which the acceleration is zero
   (vi) the time at which the acceleration is 6 m/s^2.

5. The distance, s metres, travelled in t seconds by a train after its brakes are applied is given by $s = 18t - 1.5t^2$.
   Find:
   (i) the distance travelled when $t = 2$ seconds
   (ii) the train's speed, in terms of t
   (iii) the speed of the train when $t = 4$ seconds
   (iv) the time at which the train comes to rest
   (v) the distance travelled by the train after it applied its brakes
   (vi) the constant deceleration of the train.

6. The distance, s metres, travelled by a car in t seconds after the brakes are applied is given by $s = 10t - t^2$. Show that its deceleration is constant. Find:
   (i) the speed of the car when the brakes are applied
   (ii) the distance the car travels before it stops.

7. A ball is thrown vertically up in the air. The height, h metres, reached above the ground t seconds after it was thrown is given by $h = 16t - 2t^2$.

   Find:
   (i) the height of the ball after 6 seconds
   (ii) the speed of the ball in terms of t
   (iii) the speed of the ball after 3 seconds
   (iv) the height of the ball when its speed is 12 m/s.
   (v) After how many seconds does the ball just begin to fall back downwards?
   How far above the ground is it at this time?

8. A ball is thrown vertically up in the air. Its height, h metres, above ground level varies with the time, t seconds, such that $h = 1 + 30t - 5t^2$.

   Find:
   (i) $\dfrac{dh}{dt}$, its speed
   (ii) its speed after $1\tfrac{1}{2}$ seconds
   (iii) its acceleration.
   (iv) After how many seconds does the ball just begin to fall back downwards?
   How far above the ground is it at this time?

9. The speed, v, in metres per second, of a body after t seconds is given by $v = 2t(6 - t)$.
   (i) Find the acceleration at each of the two instants when the speed is 10 m/s.
   (ii) Find the speed at the instant when the acceleration is zero.

10. The speed, v, in metres per second, of a body after t seconds is given by $v = 3t(5 - t)$.
    (i) Find the acceleration at each of the two instants when the speed is 12 m/s.
    (ii) Find the speed at the instant when the acceleration is zero.

11. An automatic valve controls the flow of gas, R cm^3/s, in an experiment. The flow of gas varies with the time, t seconds, as given by the equation $R = 8t - t^2$.

    Find:
    (i) $\dfrac{dR}{dt}$, the rate of change of R with respect to t
    (ii) the value of $\dfrac{dR}{dt}$ after 6 seconds
    (iii) the time when the rate of flow is a maximum.
    (iv) After how many seconds is the rate of flow equal to:
    (a) -4 cm^3/s (b) 2 cm^3/s?

12. The volume, V, of a certain gas is given by $V = \dfrac{20}{p}$, where p is the pressure.

    Find:
    (i) $\dfrac{dV}{dp}$, the rate of change of V with respect to p
    (ii) the value of $\dfrac{dV}{dp}$ when $p = 10$.

ANSWERS

Exercise 1.1

1. 23 2. 4 3. −3 4. −6 5. −2 6. −10 7. −12 8. 8 9. 18
10. 64 11. 20 12. −100 13. −4 14. −2 15. 6 16. −1 17. 15
18. 0 19. 0 20. 2 21. 3 22. 9 23. 6 24. 12 25. 0 26. 39
27. −39 28. 9 29. 2 30. 2 31. 4 32. 8 33. $10\frac{1}{2}$ or 10·5
34. $\frac{1}{2}$ or 0·5 35. $\frac{1}{5}$ or 0·2 36. 30 37. 68 41. (i) 17 (ii) 27
42. $x, x+1, x+2$; $3x + 3 = 3(x+1)$, which is divisible by 3

Exercise 1.2

1. $7x$ 2. $2x$ 3. $-3x$ 4. $-6x$ 5. $4x$ 6. $-7a$ 7. $-14y$ 8. $3b$ 9. $9x^2$
10. $2a^2$ 11. $3x^2$ 12. $6x^2$ 13. $-10x^2$ 14. $-12x^3$ 15. $10x^3$ 16. x^2 17. $12a^2$
18. $-15y^3$ 19. $8p^3$ 20. $x^2 + 5x + 6$ 21. $2x^2 - 3x - 20$ 22. $6x^2 - 19x + 10$ 23. a^2
24. x 25. a 26. $2ab$ 27. $2(a+4) + 8a = 10a + 8$ 28. 7 29. 3 30. 0 31. 0
32. (i) $4a$ (ii) $14x$ (iii) $4b$ 33. (i) $10x^2$ (ii) $6a^2$ (iii) $20x + 5$ or $5(4x+1)$ 34. $96y^2$

Exercise 1.3

1. 5 2. −4 3. 2 4. −3 5. 4 6. 3 7. −5 8. −3 9. 4 10. −6
11. 1 12. 2 13. 5 14. 6 15. 1 16. 5 17. 1 18. 3 19. 7 20. 4
21. 1 22. −2 23. 17 24. (i) 7 (ii) 6 (iii) 5 25. (i) $13x - 1 = 64$
    (ii) 5 (iii) 24 cm 26. (i) $(2x + 18)$ 1; $(4x + 12)$ 1 (ii) $2x + 18 = 4x + 12$ (iii) 3
27. (i) $5n + 4$ (ii) 12 28. (i) $3x + 3 = 33$ (ii) 10 (iii) 10, 11, 12 29. (i) $\frac{x}{2} + \frac{x}{5} = 70$
    (ii) 100 (iii) €20 30. 4 31. (ii) yes

Exercise 1.4

1. $x = 2, y = 1$ 2. $x = 4, y = 2$ 3. $x = 5, y = 3$ 4. $x = 5, y = 2$
5. $x = -2, y = -3$ 6. $x = 2, y = 3$ 7. $x = 7, y = 3$ 8. $x = 3, y = -1$
9. $x = 3, y = 2$ 10. $x = -1, y = 1$ 11. $x = 1, y = -4$ 12. $x = -1, y = -1$
13. $x = -2, y = -1$ 14. $x = 3, y = 4$ 15. $x = -1, y = 1$ 16. $x = 2, y = 3$
17. $x = 2, y = -3$ 18. $x = 3, y = 3$ 19. $x = 5, y = -1$ 20. $x = 2, y = -2$
21. $x = -1, y = -3$ 22. $x = 1, y = \frac{1}{3}$ 23. $x = \frac{3}{2}, y = \frac{5}{2}$ 24. $x = \frac{5}{2}, y = \frac{4}{3}$

ANSWERS

25. $x = \frac{3}{5}, y = \frac{4}{5}$ 26. $x = \frac{5}{2}, y = 1$ 27. $x = \frac{3}{2}, y = \frac{1}{2}$ 28. (i) (a) $7x + 3y = 82$
    (b) $2x + y = 24$ (ii) $x = 10, y = 4$ (iii) (a) €10   (b) €4   (iv) €124
29. A (3, −2), B (−1, 3), C (5, 2), D (1, 4)   30. (i) $x = 6, y = 2$ (ii) $x = 2, y = 1$ (iii) $x = 5, y = 3$
31. (i) $x = 50, y = 30$ (ii) $x = 40, y = 30$ 32. $5x + 3y = 49; 3x + 2y = 31; x = 5, y = 8$
33. $8x + 4y = 144; 3x + 2y = 62; x = €0·10, y = €0·16$ 34. (i) $P = 5, Q = 3, R = 4, S = 6$
    (ii) $B = 8, C = 7, D = 2, E = 5$
35. Apple = 10, Lemon = 6, Pear = 24, Grapefruit = 16, Grapes = 14 and Banana = 8

Exercise 1.5

1. $(x + 1)(x + 3)$ 2. $(x − 1)(x − 5)$ 3. $(x − 2)(x + 4)$ 4. $(x + 2)(x − 5)$
5. $(x − 1)(x − 4)$ 6. $(x + 4)(x + 5)$ 7. $(x − 3)(x + 4)$ 8. $(x + 3)(x − 5)$
9. $(x + 3)(x + 10)$ 10. $(x + 4)(x − 7)$ 11. $(2x + 3)(x + 1)$ 12. $(2x − 3)(x − 2)$
13. $(2x − 1)(x + 5)$ 14. $(2x + 3)(x − 2)$ 15. $(3x + 1)(x + 5)$ 16. $(3x − 1)(x − 7)$
17. $(3x − 2)(x + 4)$ 18. $(3x + 2)(x − 1)$ 19. $(5x + 3)(x + 1)$ 20. $(5x − 2)(x − 3)$
21. $(5x − 1)(x + 2)$ 22. $(5x + 1)(x − 4)$ 23. $(7x − 1)(x − 5)$ 24. $(11x − 2)(x + 3)$
25. $x(x + 2)$ 26. $x(x − 3)$ 27. $x(x + 4)$ 28. $x(x − 5)$
29. $x(x + 6)$ 30. $x(x − 1)$ 31. $x(x + 1)$ 32. $x(2x + 3)$ 33. $x(2x − 5)$
34. $(x − 2)(x + 2)$ 35. $(x − 4)(x + 4)$ 36. $(x − 5)(x + 5)$ 37. $(x − 8)(x + 8)$
38. $(x − 10)(x + 10)$ 39. $(x − 7)(x + 7)$
40. (i) $x^2 + 6x + 8; (x + 2)(x + 4)$ (ii) $x^2 + 5x; x(x + 5)$ (iii) $x^2 − 9; (x − 3)(x + 3)$
41. (i) $x + 4$ (ii) $2x + 1$ (iii) $x + 5$

Exercise 1.6

1. $(2x + 3)(2x + 1)$ 2. $(4x − 1)(x − 5)$ 3. $(3x + 1)(2x + 3)$ 4. $(6x − 1)(x − 2)$
5. $(3x − 2)(2x + 1)$ 6. $(8x − 1)(x − 3)$ 7. $(4x + 5)(2x − 1)$ 8. $(3x − 1)(3x − 5)$
9. $(5x − 3)(2x + 1)$

Exercise 1.7

1. 2, 3   2. −5, 4   3. 3, −7   4. 0, −3   5. 0, 5   6. 0, 8   7. ±6   8. ±4
9. ±10   10. 3, 4   11. −2, −4   12. −3, 5   13. 1, 5   14. 2, −5   15. 4, −5
16. −1, 7   17. −2, −7   18. −3, 8   19. 0, 4   20. 0, −6   21. 0, 2   22. ±3
23. ±5   24. ±1   25. $-\frac{3}{2}, -1$ 26. $\frac{3}{2}, 2$ 27. $\frac{1}{2}, -4$ 28. $\frac{2}{3}, -4$ 29. $-\frac{5}{3}, 1$
30. $\frac{1}{3}, 2$ 31. $\frac{1}{5}, -2$ 32. $\frac{2}{7}, -1$ 33. $\frac{3}{2}, -5$ 34. −3, 6   35. −2, 4   36. 3, −5
37. 0, −2   38. 0, 4   39. ±6   40. $2x^2 − x − 15; -\frac{1}{2}, 5$ 41. $-\frac{3}{2}, 2$ 42. −3, −4
43. −1, 3   44. (i) $x(x + 8) = 64$ (ii) $x^2 + 8x − 64 = 0$ 45. 2   46. (i) 5   (ii) 60

287

47. 3   **48.** 6 cm by 6 cm by 2 cm   **49.** $2x^2 + 8x = 90$; 5
50. (i) 90°; tangent and a radius are perpendicular to each other at the point of contact T   (ii) 6

Exercise 1.8

1. $2\sqrt{2}$ 2. $2\sqrt{6}$ 3. $3\sqrt{5}$ 4. $4\sqrt{2}$ 5. $3\sqrt{3}$ 6. $4\sqrt{3}$ 7. $3\sqrt{6}$ 8. $5\sqrt{5}$
9. $3\sqrt{10}$ 10. $5\sqrt{2}$ 11. (i) $1 \pm \sqrt{5}$ (ii) 1·2, 3·2   12. (i) $-1 \pm \sqrt{3}$ (ii) $-2·73, 0·73$
13. (i) $2 \pm \sqrt{5}$ (ii) $-0·24, 4·24$ 14. (i) $-3 \pm \sqrt{2}$ (ii) $-4·41, -1·59$ 15. (i) $3 \pm \sqrt{5}$
   (ii) 0·76, 5·24   16. (i) $-4 \pm \sqrt{3}$ (ii) $-5·73, -2·27$ 17. (i) $-5 \pm \sqrt{2}$
   (ii) $-6·41, -3·59$ 18. (i) $5 \pm \sqrt{7}$ (ii) 2·25, 7·65   19. (i) $-6 \pm \sqrt{3}$ (ii) $-7·33, -4·27$
20. $\dfrac{1 \pm \sqrt{3}}{2}$ 21. $\dfrac{-1 \pm \sqrt{5}}{4}$ 22. $\dfrac{-1 \pm \sqrt{2}}{3}$ 23. $-2·24, 6·24$ 24. $-0·78, 1·28$
25. $-1·84, 0·44$ 26. (i) 1·32 (ii) 0·0224   27. (i) $2 \pm \sqrt{7}$ (ii) 4·65 or $-0·65$
   (iii) $0·0225 < 0·1$ 28. (ii) 5·2   (iii) 0·16

Exercise 1.9

1. $x^2 - 5x + 6 = 0$ 2. $x^2 - x - 2 = 0$ 3. $x^2 - 3x - 10 = 0$ 4. $x^2 - 3x - 4 = 0$
5. $x^2 + 5x + 6 = 0$ 6. $x^2 - 9x + 20 = 0$ 7. $x^2 - x - 12 = 0$ 8. $x^2 + 5x - 24 = 0$
9. $x^2 - 9 = 0$ 10. $x^2 - 4 = 0$ 11. $x^2 + 2x = 0$ 12. $x^2 - 5x = 0$ 13. $x^2 - 1 = 0$
14. $2x^2 - 7x + 3 = 0$ 15. $3x^2 - 5x - 2 = 0$ 16. $2x^2 + 5x - 3 = 0$ 17. $2x^2 - 3x - 5 = 0$
18. $6x^2 - 5x + 1 = 0$ 19. $9x^2 + 3x - 2 = 0$ 20. $8x^2 - 10x + 3 = 0$ 21. $x^2 - 2x - 15 = 0$
22. $10x^2 + x - 2 = 0$

Exercise 1.10

1. (i) $x = 2$ and $y = -3$ or $x = 4$ and $y = -1$ (ii) $x = -1$ and $y = -1$ or $x = 3$ and $y = 11$
2. (i) $(-2, 1)$ and $(5, 8)$ (ii) $(-4, -11)$ and $(3, 3)$
3. $(3, 5)$, one point of intersection, therefore line is a tangent
4. $x = 3$ and $y = 2$ or $x = 2$ and $y = 3$ 5. $x = 2$ and $y = 1$ or $x = -1$ and $y = -2$
6. $x = 1$ and $y = 4$ or $x = 2$ and $y = 2$ 7. $x^2 + 6x + 9$ 8. $x^2 - 4x + 4$ 9. $y^2 - 2y + 1$
10. $y^2 + 6y + 9$ 11. $4 + 4x + x^2$ 12. $1 - 4x + 4x^2$ 13. $9 - 12x + 4x^2$
14. $9 - 30y + 25y^2$ 15. $x = -1$ and $y = 4$ or $x = 4$ and $y = -1$
16. $x = -3$ and $y = -4$ or $x = 4$ and $y = 3$ 17. $x = -1$ and $y = 0$ or $x = 0$ and $y = 1$
18. $x = 4$ and $y = 2$ or $x = -4$ and $y = -2$ 19. $x = -3$ and $y = -4$ or $x = 5$ and $y = 0$
20. $x = 2$ and $y = 3$ or $x = 3$ and $y = 2$ 21. $x = 5$ and $y = 1$ or $x = 0$ and $y = -4$
22. $x = 1$ and $y = 2$ 23. $x = -1$ and $y = 3$ or $x = 2$ and $y = -3$
24. (i) $x = 5$ and $y = 0$ or $x = 3$ and $y = 4$ (ii) 125, 91   25. (ii) 4 cm by 3 cm

26. (ii) $x = 3$ and $y = 9$; the lengths of the 3 sides are 6 cm, 8 cm and 10 cm
27. (i) $A(1, 2)$ and $B(8, 9)$ (ii) $\sqrt{98}$ or $7\sqrt{2}$ 28. $P(-6, 1)$ and $Q(2, 3)$
29. (i) $(-2, 10)$ and $(2, 10)$ (ii) $(-5, 16)$ and $(5, 16)$

Exercise 1.11

1. $x \geq 3$ 2. $x \leq 4$ 3. $x > 2$ 4. $x < 2$ 5. $x \leq 3$ 6. $x \geq -5$ 7. $x < -2$
8. $x \leq 1$ 9. $x \geq 1$ 10. $x \leq 3$ 11. $x \leq 2$ 12. $x \leq 4$ 13. $x > 5$
14. $x < 2$ 15. $2 \leq x \leq 5$ 16. $-1 \leq x < 4$ 17. $-2 \leq x < 3$ 18. $-3 < x \leq 4$
19. (i) $x \geq 3$ (ii) $x \leq 5; 3 \leq x \leq 5$ 20. (i) $x \leq 4$ (ii) $x \geq -2; -2 \leq x \leq 4$
21. (i) $x \leq 6$ (ii) $x \geq -4$ (iii) $-4 \leq x \leq 6$ 22. (i) $x \geq -6$ (ii) $x \leq 1$ (iii) $-6 \leq x \leq 1$
23. (i) $x \leq 2$ (ii) $x > -3$ (iii) $-2, -1, 0, 1, 2, 3$ 24. (i) $x \leq 1$ (ii) $x \leq 3$ (iii) $2, 3$
25. 3   26. $x \geq 5, x \leq 4$; nothing in common, no intersection   27. (i) 2, 4   (ii) 1
(iii) 1, 4, 9, 16   (iv) 2, 3, 5, 7   28. 3, 4, 5   29. $x > 1$ 30. (i) D   (ii) A   (iii) B
(iv) C   (v) A   (vi) C   (vii) C   (viii) A   31. (ii) $2 < x < 7$ 32. (i) 4, 5, 6, 7 or 8
(ii) yes, Bernadette and Catherine, both aged 14 or Bernadette and Dermot, both aged 22   33. $(10, 7)$

Exercise 1.12

1. $\dfrac{b+c}{2}$ 2. $\dfrac{r-q}{3}$ 3. $\dfrac{c+d}{b}$ 4. $\dfrac{v-u}{t}$ 5. $\dfrac{5c-3a}{2}$ 6. $\dfrac{4p+2r}{3}$ 7. $\dfrac{2b+c}{2}$
8. $\dfrac{ac+d}{a}$ 9. $\dfrac{w-xz}{x}$ 10. $2b$ 11. $2a - 2c$ 12. $r - 3s$ 13. $\dfrac{6c-2b}{3}$ 14. $2r - p$
15. $3r + q$ 16. $\dfrac{b-3a}{2}$ 17. $zy - xy$ 18. $\dfrac{sr-2pr}{3}$ 19. $5q + 3r$ 20. $\dfrac{p+r}{s}$
21. $\dfrac{2a-3c}{d}$ 22. $\dfrac{2c-3b}{9}$ 23. $\dfrac{v^2-u^2}{2s}$ 24. $\dfrac{b-4a}{8}$ 25. $\dfrac{2s}{t^2}$ 26. $\dfrac{3v}{\pi r^2}$
27. $\dfrac{2s-2ut}{t^2}$ 28. $\dfrac{1}{r-t}$ 29. $\dfrac{t}{r-p}$ 30. $\dfrac{y}{x-w}$ 31. (i) $\dfrac{v-u}{a}$ (ii) 40   32. (i) $\dfrac{2A}{a+b}$ (ii) 12
33. (i) $P = 2l + 2w$ (ii) $w = \dfrac{P-2l}{2}$ (iii) $A = lw$ (iv) $w = \dfrac{A}{l}$ (v) $A = \dfrac{Pl - 2l^2}{2}$
34. (i) $2x + y = 18$ (ii) $y = 18 - 2x$ (iii) $x(18 - 2x) = 40$ (iv) $x = 4$ and $y = 10$
35. (i) (a) $\dfrac{9c+160}{5}$ (b) $\dfrac{9k-113}{5}$ (ii) (a) 50   (b) -1

Exercise 1.13

1. 2^7 2. 5^9 3. 7^4 4. 3^7 5. 3^2 6. 2^3 7. 3^{-2} 8. 5^{-3} 9. 3^8 10. 5^{10}
11. 5^2 12. $3^{\frac{3}{2}}$ 13. $2^{\frac{3}{2}}$ 14. $3^{\frac{7}{2}}$ 15. $2^{-\frac{5}{2}}$ 16. $5^{-\frac{3}{2}}$ 17. a^2 18. a^3
19. a^2 20. a^4 21. 2^2 22. 5^2 23. 6^2 24. 3^3 25. 2^4 or 4^2 26. 7^2
27. 2^5 28. 3^4 or 9^2 29. 5^3 30. 2^7 31. 3^5 32. 6^3 33. 8   34. 9

35. 64 36. 25 37. 36 38. 125 39. 81 40. 125 41. 49 42. 16 43. 216
44. 1 45. $\frac{1}{3}$ 46. $\frac{1}{16}$ 47. $\frac{1}{125}$ 48. $\frac{1}{32}$ 49. $\frac{1}{9}$ 50. $\frac{1}{1,000}$ 51. 3 52. 5
53. 2 54. 4 55. 2 56. 6 57. 8 58. $\frac{1}{8}$ 59. 16 60. $\frac{1}{16}$ 61. 9 62. $\frac{1}{9}$
63. 2^6 or 4^3 or 8^2 64. 3^4 or 9^2 65. 2^3 66. 3^2 67. 5^2 68. 7^2 69. 2^4 70. 5^3
71. $2^{\frac{1}{2}}$ 72. $3^{\frac{1}{2}}$ 73. $5^{\frac{1}{2}}$ 74. $7^{\frac{1}{2}}$ 75. $2^{\frac{3}{2}}$ 76. 2^3 77. $2^{\frac{3}{2}}$ 78. $3^{\frac{3}{2}}$ 79. $5^{\frac{3}{2}}$
80. $5^{\frac{3}{2}}$ 81. $7^{-\frac{1}{2}}$ 82. $5^{\frac{5}{2}}$ 83. 2^3 84. 3^{-1} 85. 5^{-3} 86. 2^1 or 2 87. 5^3
88. 3^{-3} 90. 9^{-1} and 3^{-2}, both are equal to $\frac{1}{9}$ 91. (i) $(x^2 + x)$ m^2 (ii) $x^2 + x = 42$; 6
92. (i) $x^2 - x$ (ii) 3 93. (i) (a) a (b) b (c) x (d) $x + 3$ (ii) (a) $(x + 1)^2$ (b) $(x + 2)^2$
(iii) $2x + 3$ (iv) 3 94. (i) $x^{\frac{1}{2}}$ (ii) $x^{\frac{1}{3}}$ (iii) $x^{\frac{2}{3}}$ (iv) $x^{\frac{3}{4}}$

Exercise 1.14

1. 4 2. 3 3. 2 4. 1 5. 2 6. 1 7. 6 8. 2 9. 1 10. 4 11. 2
12. 1 13. $\frac{3}{2}$ 14. $\frac{3}{4}$ 15. $\frac{1}{4}$ 16. 2 17. -2 18. -3 19. -2 20. -1
21. $-\frac{5}{2}$ 22. $\frac{1}{2}$ 23. $\frac{1}{2}$ 24. -10 25. 4^2; $\frac{3}{2}$ 26. 2^4; 3 27. $-1, 2$ 28. $2^{\frac{3}{2}}$; $\frac{1}{4}$
29. (i) (a) $2^{\frac{1}{2}}$ (b) 2^3 (c) 2^4 (d) $2^{\frac{7}{2}}$ (ii) (a) 1 (b) $\frac{3}{2}$ (c) 2 30. (i) (a) $3^{\frac{1}{2}}$ (b) $3^{-\frac{1}{2}}$
(c) 3^4 (d) $3^{\frac{7}{2}}$ (ii) (a) $\frac{1}{4}$ (b) $\frac{1}{4}$ (c) $\frac{11}{2}$ 31. (i) (a) 5^3 (b) $5^{\frac{1}{2}}$ (c) $5^{\frac{5}{2}}$ (d) 5^5
(ii) (a) $-\frac{1}{6}$ (b) $-\frac{1}{4}$ (c) 3 32. (i) $x + 1$ (ii) 4 33. (i) $\frac{1}{2}bh$ (ii) 9 34. (i) 2^{k+2}
(ii) (a) 3 (b) $\frac{1}{2}$ 35. (i) $\dfrac{8a}{c + 5}$ (ii) $\sqrt{2}$ or $2^{\frac{1}{2}}$

Exercise 2.1

1. $\frac{1}{6}$; $16\frac{2}{3}$% 2. (i) $\frac{1}{10}$ (ii) €120
3. (i) (a) e: $\frac{1}{6}$ or $16\frac{2}{3}$%; a: $\frac{1}{18}$ or $5\frac{5}{9}$%; t: $\frac{1}{9}$ or $11\frac{1}{9}$% (b) e: more; a: less; t: more
   (ii) (a) e: $\frac{7}{66}$ or 10·6%; a: $\frac{4}{66}$ or 6·06%; t: $\frac{7}{66}$ or 10·6% (b) e: less; a: less; t: more
4. (i) $\frac{5}{8}$ (ii) 144,000 5. €550 6. 27·2 km

Exercise 2.2

1. (i) €56; €24 (ii) 250 g; 200 g 2. (i) €120; €160; €200 (ii) €1,000; €1,600; €1,400
3. (i) 119 g; 34 g; 85 g (ii) 72 cm; 54 cm; 36 cm 4. (i) €126; €168; €210
   (ii) 48 cm; 120 cm; 168 cm 5. (i) €102; €119; €153 (ii) 960 g; 120 g; 480 g 6. 3 : 4
7. 1 : 3 8. 2 : 1 : 4 9. 3 : 2 : 4 10. (i) €28; €14 (ii) 56 g; 224 g 11. (i) €60; €120; €30
   (ii) 39 cm; 156 cm; 390 cm 12. (i) 252 g; 168 g; 126 g (ii) €240; €320; €360
13. Team A received €15,750; Team B received €12,250 14. €178,800
15. (i) 5 : 9 : 6 (ii) David €125, Eric €225, Fred €150 16. €120 17. 100 cm 18. 162 cm
19. (i) €30 (ii) €165 20. €10,160 21. 35 cm 22. €9,500 23. $k = 5$
24. (i) Roy: €320, Sam €120 (ii) 5 : 3

ANSWERS

Exercise 2.3
1. (i) €9·60 (ii) €25·92 (iii) €26·04   2. €4,573·80   3. (i) €288 (ii) (a) €244·80 (b) 2%
4. 20%   5. 60%   6. €82   7. €10·08   8. (i) €1,200 (ii) €1,452   9. €72·80
10. €3,267   11. 700   12. 160   13. €14,000   14. €232,000   15. €1,560
16. €350,000   17. 2·8 litres   18. (i) 400 (ii) 700   19. €20   20. (i) €1,533 (ii) 18
21. (i) €260 (ii) €82·80 (iii) 42%   22. No   23. Better offer   24. €6·35; €8·25; €15·88
25. Less than

Exercise 2.4
1. 2; $\frac{1}{6}$; 16·7%   2. 3; $\frac{3}{43}$; 7·0%   3. 4; $\frac{4}{136}$; 2·9%   4. 0·2; $\frac{1}{24}$; 4·2%   5. 0·3; $\frac{1}{19}$; 5·3%
6. 10; $\frac{1}{39}$; 2·6%   7. (i) 0·15 m (ii) 9·1%   8. (i) 2·5 kg (ii) 3·7%   9. (i) 5 km
   (ii) $\frac{5}{105}$ or $\frac{1}{21}$ (iii) 4·8%   10. 5%   11. 5·44%   12. 2·3%   13. (i) $\frac{1}{10}$
   (ii) 1·64%   14. (i) 1,584 cm³ (ii) 4·5%   15. 10·3%   16. (i) 4% (ii) 0·8%
17. (i) 1% (ii) 6,187·5 kg (iii) 12·375 tins (iv) €62·50

Exercise 2.5
1. 10; 14   2. 115; 125   3. 75%; 85%   4. 3·995 m; 4·005 m   5. €2,750; €3,650
6. −2; 8   7. 150 ± 3 sweets   8. (i) 0·2 m or 20 cm
9. (i) 30,000 cm³ (ii) 29,601 cm³; 30,401 cm³ (iii) 6·4 kg

Exercise 2.6
1. (i) $257·50 (ii) €600   2. €50   3. (i) ¥72,960 (ii) €380   4. South Africa; €30
5. (i) R11,000 (ii) R750   6. (i) €8·10 (ii) $1\frac{1}{2}$%   7. €3,587·50   8. (i) €48 (ii) 2%
9. $2\frac{1}{2}$%   10. $880   11. (i) 30% (ii) 24·8%

Exercise 2.7
1. €1,996·80   2. €2,173·50   3. €2,837·25   4. €6,492·80   5. €248·25
6. €306·04   7. €1,664·39   8. €6,151·25   9. €368   10. €1,755·52   11. €8,234·85

Exercise 2.8
1. €24,812·48   2. €26,981·76   3. €12,624·60   4. 3%   5. 4%   6. $3\frac{1}{2}$%
7. $2\frac{1}{2}$%   8. (i) 5% (ii) $4\frac{1}{2}$%   9. (i) €62,400 (ii) 3%   10. (i) €43,056 (ii) $2\frac{1}{2}$%
11. (i) €22,950 (ii) 3%   12. (i) €10,868 (ii) 3%   13. (i) €68,500 (ii) $3\frac{1}{2}$%
14. (i) €33,075 (ii) €8,075   15. €7,120   16. w = €5,600   17. w = €4,450

Exercise 2.9
1. €18,423·75   2. €36,085·50   3. €25,710

4. (i)

Years after purchase	0	1	2	3	4
Value of car (€)	50,000	40,000	32,000	25,600	20,480

(ii)
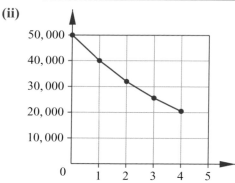

5. (i) 8%   (ii) €17,910

Exercise 2.10
1. €29,549·11   2. €19,661·94   3. €45,551·48   4. €29,255·87   5. €5,955·08
6. €15,315·38   7. €7,813·56   8. €4,415·25   9. (i) (a) €20   (b) €20·50
(ii) (a) 10%   (b) 10·25%   (iii) €20·76   10. (i) €110   (ii) €112·36   (iii) €112·62
(iv) 10%; 12·36%; 12·62%   11. (i) €1,195·62   (ii) 19·6%
12. (i) $P = \dfrac{f}{(1+i)^t}$ (ii) €50,000   13. €200,000   14. (i) 1·046%   (ii) €1,882·32   16. 1·02%

Exercise 2.11
1. (i) 30 mins   (ii) 36 mins   (iii) 6 mins   (iv) 42 mins   (v) 24 mins
2. (i) 1 hr 12 mins   (ii) 2 hrs 45 mins   (iii) 3 hrs 21 mins   (iv) 1 hr 20 mins   (v) 2 hrs 27 mins
3. (i) $\frac{1}{4}$ hr or 0·25 hr   (ii) $\frac{1}{3}$ hr or $0·\dot{3}$ hr   (iii) $1\frac{1}{2}$ hrs or 1·5 hrs   (iv) $2\frac{3}{4}$ hrs or 2·75 hrs
(v) $3\frac{2}{5}$ hrs or 3·4 hrs   4. 80 km/h   5. 1 hr 45 mins   6. 150 km   7. 15:29
8. (i) 1 hr 15 mins   (ii) 12:30   (iii) €7·75   9. 64 km/h   10. 4 m/s   11. 18:30
12. (i) $4\frac{1}{4}$ hrs or 4·25 hrs   (ii) 60 km/hr   13. 72 km/h   14. 91 km   15. 08:35
16. (i) 18 m/s   (ii) 64·8 km/h   17. 21 mins   18. 14·25 km or 14,250 m   19. 09:05
20. 200 m race   21. 32 km/h   22. 29·4 km   23. (i) 8·67 m/s; 8·82 m/s; 8·88 m/s; 8·92 m/s
(ii) 8·82 m/s   24. 54 km/h   25. (i) 280 km   (ii) 5 hrs   (iii) 56 km/h
26. (i) 6 hrs   (ii) 462 km   (iii) 77 km/h   27. (i) 4 hrs   (ii) 260 km   (iii) 65 km/h
28. 90 km/h   29. (i) 504 km   (ii) 8 hrs   (iii) 63 km/h   30. (i) 55 km/h   (ii) 66 km/h
(iii) 11 hrs; 660 km   (iv) 60 km/h

ANSWERS

Exercise 2.12
1. €160 2. €350 3. €920 4. €260 5. €620 6. €0 7. €9,020 8. €5,420
9. €190 10. €150 11. €359·28 12. €1,278·80 13. €2,118·80 14. €5,618·80
15. €16,118·80 16. €80·20 17. (i) Alex: €4,000; Bruce: €4,550 (ii) Alex: €0; Bruce: €91
(iv) €18

Exercise 2.13
1. (i) €11,100 (ii) €7,300 2. (i) €9,960 (ii) €7,470 3. (i) €9,370 (ii) €6,130
4. (i) €11,876·50 (ii) €8,726·50 5. 20% 6. 18% 7. (i) €19,000 (ii) €48,000
8. (i) €23,500 (ii) €52,000

Exercise 2.14
1. Gross tax: €2,000; USC: €200; Take-home pay: €9,300
2. Gross tax: €11,600; USC: €2,120; Take-home pay: €35,480
3. Gross tax: €20,400; USC: €4,220; Take-home pay: €55,380
4. Gross tax: €41,600; USC: €6,620; Take-home pay: €76,580
5. Gross tax: €3,150; USC: €399·28; Take-home pay: €14,600·72
6. Gross tax: €9,980; USC: €1,978·80; Take-home pay: €31,041·20
7. Gross tax: €55,900; USC: €9,818·80; Take-home pay: €90,381·20
8. Gross tax: €117,400; USC: €20,318·80; Take-home pay: €169,781·20

Exercise 2.15
1. 8×10^3 2. $5·4 \times 10^4$ 3. $3·47 \times 10^5$ 4. $4·7 \times 10^2$ 5. $2·9 \times 10^3$ 6. $3·4 \times 10^6$
7. $3·94 \times 10^2$ 8. $3·9 \times 10^1$ 9. 6×10^{-3} 10. 9×10^{-4} 11. $5·2 \times 10^{-2}$
12. $4·32 \times 10^{-4}$ 13. $4·2 \times 10^3$ 14. $7·8 \times 10^4$ 15. $3·2 \times 10^{-2}$ 16. $4·5 \times 10^{-3}$
17. $n = 3$ 18. $n = 4$ 19. $n = -2$

Exercise 2.16
1. $3·2 \times 10^3$ 2. $2·8 \times 10^6$ 3. $5·2 \times 10^5$ 4. $2·3 \times 10^4$ 5. $7·8 \times 10^{-3}$
6. $2·94 \times 10^{-4}$ 7. $7·2 \times 10^7$ 8. $3·6 \times 10^7$ 9. $7·48 \times 10^5$ 10. $9·54 \times 10^6$
11. $2·3 \times 10^3$ 12. $1·4 \times 10^5$ 13. $2·4 \times 10^3$ 14. $5·8 \times 10^3$ 15. $1·5 \times 10^5$
16. $2·8 \times 10^{-4}$ 17. $3·51 \times 10^6$ 18. $5·4 \times 10^3$ 19. 3×10^3 20. 5×10^2
21. $1·52 \times 10^2$ 22. $2·8 \times 10^3$ 23. 0·078; less 24. 0·2; greater 25. $k = 20$

Exercise 2.17
1. (i) €25,871·87 (ii) €18,814·83 2. 1,398 3. (i) €574,349·12 (ii) €768,608·68
(iii) 40 years 5. €186 6. (ii) €2,665·84 (iii) 2·8%
7. (i) $11\frac{1}{2}$ hours (ii) €86·25 8. (i) €390 (ii) 15–21 days before departure (iii) €1,260
9. (i) €48·40 (ii) €58·56 10. Brand C 11. $2·8 \times 10^{-6}$ mm^2 12. (i) $1·5 \times 10^5$ m
(ii) 5×10^{-11} seconds 13. $7·6 \times 10^{-3}$ 14. (i) $1·44 \times 10^6$ (ii) 3,550

15. (i) 1.392×10^8 km (ii) 2.19×10^{19} km (iii) 4.99×10^3 16. 8.8×10^6 tonnes
17. 2×10^{-7} 18. 11 people per square km 19. (i) 1.2×10^{-2} cm (ii) 3.4992×10^9
(iii) 420 km 20. 1.68×10^7 21. 320 times 22. 6.37×10^3 km 23. 2.05×10^7
24. 200 25. $k = 3 \times 10^4$ 26. 3.64×10^{14} m² 27. 2.5×10^4 cm

Exercise 3.1
1. $4i$ 2. $3i$ 3. $2i$ 4. $5i$ 5. $8i$ 6. $10i$ 7. $7i$ 8. $12i$

Exercise 3.2
1. $R = 5, I = 3$ 2. $R = 2, I = 5$ 3. $R = 6, I = 7$ 4. $R = 5, I = 4$
5. $R = 2, I = -7$ 6. $R = -4, I = 6$ 7. $R = -3, I = -5$ 8. $R = -9, I = 8$
9. $R = 2, I = 1$ 10. $R = 3, I = -1$ 11. $R = -5, I = 1$ 12. $R = -1, I = -1$
13. $R = 6, I = 0$ 14. $R = 0, I = 2$ 15. $R = -2, I = 0$ 16. $R = 0, I = -5$

Exercise 3.3
1. $6 + 5i$ 2. $3 + 4i$ 3. $6 + 7i$ 4. $-2 + 3i$ 5. $2 + 4i$ 6. $-3 + 6i$
7. $14 + 5i$ 8. $17 + 16i$ 9. $0 - 3i$ 10. $15 + 0i$ 11. $5 - i$ 12. $3 + 2i$
13. $4 - i$ 14. $2 + 0i$ 15. $3 + 0i$ 16. $4 - 2i$ 17. $5 + 3i$ 18. $5 + 0i$
19. $0 + 7i$ 20. $3 - 2i$ 21. $5 - 5i$ 22. $2 + 5i$ 23. $5 + i$ 24. $8 + 4i$
25. $-4 - 32i$

Exercise 3.4
1. $4 + 6i$ 2. $-6 + 12i$ 3. $-2 - i$ 4. $4 + 3i$ 5. $1 + 2i$ 6. $3 - i$ 7. $5 + 14i$
8. $5 + 5i$ 9. $16 - 11i$ 10. $21 - 22i$ 11. $-10 - 5i$ 12. $13 + 0i$ 13. $3 + 4i$
14. $5 - 12i$ 15. $21 + 20i$ 16. $5 + i$ 17. $-5 + 12i$ 18. $-24 - 10i$
19. $-2 + 0i$ 20. $-3 + i$ 21. $-8 + i$ 22. $-1 + 5i$ 23. $-2 + 0i$

Exercise 3.5
1. $3 - 2i$ 2. $4 + 3i$ 3. $-2 - 6i$ 4. $-3 + 7i$ 5. $1 + 5i$ 6. $-1 - 3i$
7. $-4 + 5i$ 8. $-2 - 3i$ 9. $7 - 3i$ 10. $3 - i$ 11. $-8 - 6i$ 12. (i) 8   (ii) $10i$
(iii) 41   13. (i) 6   (ii) $-4i$ (iii) 13   14. (i) -8 (ii) $4i$ (iii) 20   15. (i) -2
(ii) $-2i$ (iii) 2   16. 2   17. 6   18. 17   21. $\sqrt{10}$

Exercise 3.6
1. $2 + 2i$ 2. $2 + 3i$ 3. $2 + i$ 4. $-2 + i$ 5. $5 + 2i$ 6. $3 - 4i$ 7. $3 - 5i$
8. $2 - 3i$ 9. $4 + 0i$ 10. $0 - i$ 11. $\frac{1}{2} + \frac{3}{2}i$ 12. $\frac{6}{5} - \frac{8}{5}i$ 13. $5 + i$ 14. $3 - 5i$

15. (i) $2 + i$ (ii) $3 + i$ **16.** (i) $4 + 0i$ (ii) 6 **17.** -1 **18.** (i) yes (ii) $3 + 4i$; 5
19. real $= 2$ and imaginary $= 3$

Exercise 3.7

1. $x = 5, y = 3$ **2.** $x = 6, y = 3$ **3.** $x = 2, y = 1$ **4.** $x = 3, y = 5$ **5.** $x = 2, y = 3$
6. $x = 1, y = 4$ **7.** $x = 2, y = -1$ **8.** $x = -4, y = 3$ **9.** $x = -2, y = 5$
10. $x = 8, y = -2$ **11.** $k = 3, l = 2$ **12.** $p = 4, q = -1$ **13.** $k = 2, t = 2$
14. $k = 12$ **15.** $k = -1, k = -5$ **16.** $p = 1, q = 2$ **17.** (i) $3 - 4i$ (ii) $k = -1, t = 2$
18. (ii) $k = -1, t = -3$ **19.** (ii) $p = 3, q = -1$ **20.** (i) $2 + 2i$ (ii) $a = 1, b = -4$
21. $s = 36, t = 2$ **22.** $x = 1, y = 3$ **23.** (i) $\frac{11 + 10i}{17}$ (ii) $p = 11, q = 5$ **24.** $a = 3, b = 1$

Exercise 3.8

1. $3 \pm 2i$ **2.** $1 \pm 3i$ **3.** $-2 \pm i$ **4.** $5 \pm 3i$ **5.** $-2 \pm 3i$ **6.** $5 \pm 4i$ **7.** $-1 \pm i$
8. $1 \pm 2i$ **9.** $-4 \pm i$ **10.** $\pm 2i$ **11.** $\pm 5i$ **12.** $\pm 3i$ **13.** $\frac{1}{2} \pm \frac{1}{2}i$ **14.** $\frac{3}{2} \pm \frac{1}{2}i$
15. $1 \pm \frac{1}{2}i$ **16.** $1 + 2i$ **17.** $2 + 5i$ **18.** $4 + 3i$ **19.** $-1 + 2i$ **20.** $-6 + i$
21. (i) $p = 2, k = 3$ (ii) yes **22.** (i) $-2 + i$ (ii) $-2 - i$ **23.** $3 + 5i$ **24.** $x^2 + 4x + 5 = 0$
25. $x^2 - 2x + 2 = 0$ **26.** $x^2 + 2x + 10 = 0$ **27.** $x^2 - 6x + 34 = 0$ **28.** $x^2 - 16 = 0$
29. $x^2 - 1 = 0$ **30.** $p = -6, q = 34$ **31.** $a = -14, b = 50$ **32.** $m = 6, n = 18$
33. $k = 26$

Exercise 3.9

1. $x = 2 \pm 3i$ or $b^2 - 4ac = -36 < 0$ **2.** $x = -3 \pm 4i$ or $y = -2 \pm 4i$ or $b^2 - 4ac = -64 < 0$
3. $x = 1 \pm 2i$ or $y = -2 \pm 2i$ or $b^2 - 4ac = -16 < 0$ **4.** $x = 4 \pm 5i$ or $b^2 - 4ac = -100 < 0$
5. $x = \pm 3i$ or $b^2 - 4ac = -36 < 0$ **6.** $x = \pm 2i$ or $b^2 - 4ac = -16 < 0$
7. $x = -2 \pm 5i$ or $y = 2 \pm 5i$ or $b^2 - 4ac = -100 < 0$ **8.** $x = 3 \pm 5i$ or $y = 3 \pm 5i$ or $b^2 - 4ac = -100 < 0$
9. $x = 1 \pm i$ or $y = 1 \pm i$ or $b^2 - 4ac = -4 < 0$ **10.** $x = 1 \pm 4i$ or $y = \pm 4i$ or $b^2 - 4ac = -64 < 0$
11. $x = -2 \pm i$ or $y = 2 \pm i$ or $b^2 - 4ac = -4 < 0$ **12.** $x = -3 \pm i$ or $y = 3 \pm i$ or $b^2 - 4ac = -4 < 0$
13. $x = 4 \pm i$ or $y = 4 \pm i$ or $b^2 - 4ac = -4 < 0$ **14.** $x = 5 \pm 2i$ or $y = 5 \pm 2i$ or $b^2 - 4ac = -16 < 0$
15. $x = -2 \pm 5i$ or $y = 1 \pm 5i$ or $b^2 - 4ac = -100 < 0$

16. (ii)

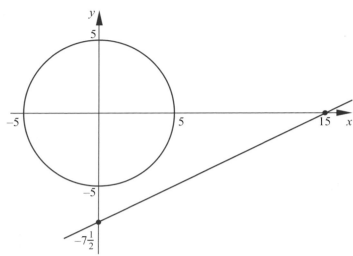

(iii) $x = -6 \pm 2i$ or $b^2 - 4ac = -16 < 0$ 17. $x = 2 + 3i$ and $y = 2 - 3i$ or $x = 2 - 3i$ and $y = 2 + 3i$
18. $a = 5 + 2i$ and $b = 5 - 2i$ or $a = 5 - 2i$ and $b = 5 + 2i$

Exercise 3.10

1.

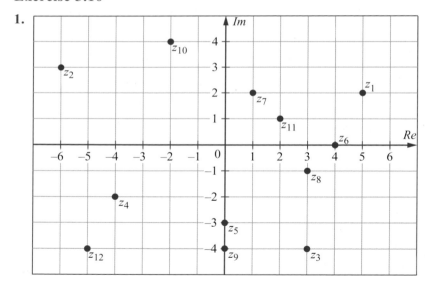

3. (i) -6 (ii) 2

Exercise 3.11

1. 5   2. 10   3. 13   4. 17   5. 26   6. 25   7. 29   8. 41   9. 3   10. $\sqrt{5}$
11. $\sqrt{61}$ 12. 13   13. 10   14. 1   15. $\sqrt{2}$ 16. $\sqrt{13}$ 18. $\sqrt{10}$
20. $1 + i$; 2   21. no   22. no   23. no   24. (i) $5 - 5i$ 25. (i) $-21 + 20i$

ANSWERS

26. **(i)** $22 + 4i$; $-2 + i$ 27. $\frac{13}{10}$ or 1.3 28. **(i)** 5 **(ii)** $p = 3, q = 2$ **(iii)** $s = \frac{3}{25}, t = \frac{4}{25}$
29. w, as $\sqrt{164} < 13$ 30. **(i)** $x = 4, y = 3$ 31. **(i)** $3 - 4i$ **(iii)** $h = 2, k = -\frac{2}{5}$
32. $a = 1, b = -2$ 34. **(ii)** ± 4 35. ± 6 36. ± 2 37. ± 5 38. ± 4

Exercise 3.12
1. -1 2. $-i$ 3. 1 4. i 5. -1 6. $-i$ 7. 1 8. -1 9. i 10. 1
11. -1 12. $-i$ 13. 2 14. $-3i$ 15. **(i)** $-1 + 5i$ **(ii)** $-2 + 3i$ **(iii)** $6 - 2i$
18. **(i)** -1 **(ii)** $-i$ **(iii)** 1 **(iv)** -1 19. **(i)** -1 **(ii)** $-i$ **(iii)** 1 **(iv)** 1

Exercise 3.13
1. yes, both the same distance, 5, from the origin. 2. **(ii)** same as z **(iii)** central symmetry in the origin, $O(0, 0)$ **(iv)** z **(v)** rotation of $90°$ and $450°$, about the origin, $O(0, 0)$ maps (moves) a point to the position **(vi)** rotate by $90°$ and double its distance from the origin (in a straight line from the origin) or vice versa **(vii)** $5 > \sqrt{5}$
3. **(i)** $u = 4 + 2i$, $v = 2 + i$ **(ii)** $k = 2$ **(iii)** $l = \frac{1}{2}$ **(iv)** on the same straight through the origin with $2i^3 v$ twice as far as $-iv$ from the origin, $O(0, 0)$; $(2 - 4i = 2(1 - 2i))$
4. **(ii)** rotation of $360°$, about the origin, maps (moves) a point to it original position; **(iii)** yes; rotation of $180° = $ rotation of $-180°$, about the origin, $O(0, 0)$ **(iv)** rotate by $90°$ and treble its distance from the origin (in a straight line from the origin) or vice versa
5. **(iii)** no; $iz_1 = -4 + 2i$, $z_2 = 2 - 4i$ **(iv)** axial symmetry in the real axis **(v)** yes
6. **(iv)** circle; centre $= (2, 1)$ and radius $= 5$ **(v)** yes **(vi)** $5 + 5i$ 7. **(i)** $1 + 2i$
8. **(i)** $z_1 = 2 + i$; $z_2 = 1 + 2i$; $z_3 = 4 + 2i$; $z_4 = 3 + 4i$ **(ii)** $k = 2$
9. **(i)** $z^2 = 0 + 2i$, $z^3 = -2 + 2i$ **(iii)**

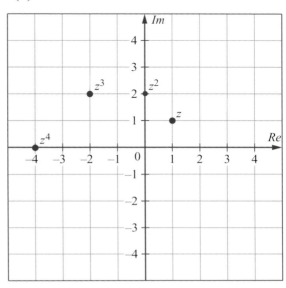

   (iv) Each time z is multiplied by itself it rotates by $45°$ about the origin and it also moves further away from the origin by $\sqrt{2}$ each time. If we keep multiplying z by itself and join the points the curve will look like a spiral. **(v)** $z^8 = 16$, $z^{12} = -64$, $z^{16} = 256$ **(vi)** $z^{40} = (z^4)^{10} = (-4)^{10} = 1,048,572$ positive number. Alternatively, (-4) to an even exponent (power) will always be positive **(vii)** $z^{40} = (-2)^{20}$
   (viii) $z^{41} = 1,048,576(1 + i)$ or $1,048,576 + 1,048,576i$
   (ix) $(\sqrt{2})^{41}$ or $2^{20}\sqrt{2}$ or $1,048,576\sqrt{2}$ or $1,482,910.4$

Exercise 4.1

1. (i) 64 cm (ii) 240 cm^2 2. (i) 60 cm (ii) 225 cm^2 3. (i) 24 cm (ii) 24 cm^2
4. (i) 94·2 cm (ii) 706·5 cm^2 5. (i) 35·7 cm (ii) 78·5 cm^2 6. (i) 52·56 cm (ii) 125·6 cm^2
7. (i) 56 cm (ii) 144 cm^2 8. (i) 68 cm (ii) 290 cm^2 9. (i) 80 cm (ii) 280 cm^2
10. 24 m^2 11. 276 m^2 12. 60 cm^2 13. 3,000 cm^2 14. (i) $10\sqrt{3}$ (ii) 17·3 cm^2
15. (i) $16\sqrt{3}$ (ii) 27·7 cm^2 16. (i) $6\sqrt{5}$ (ii) 13·4 cm^2 17. 24 cm^2 18. 19·5 cm^2
19. 23·4 cm^2 20. 108 cm^2 21. 251·3 cm^2 22. 86 cm^2
23. (i) 16 cm (ii) 128π cm^2 (iii) 50% 24. (i) 420 cm^2 (ii) 29 cm
25. (i) 72 cm (ii) 324 cm^2 26. 513π cm^2 27. 71·5 cm 28. 19 cm 29. 400 m
30. (i) 100 m (ii) D (iii) 7·3 m/sec 31. $540 - 80\pi$ cm^2

Exercise 4.2

1. (i) 23 cm (ii) 483 cm^2 2. (i) 8 cm (ii) 96 cm 3. (i) 81 cm^2 (ii) 20 cm
4. 8 cm 5. 15 m 6. 8 cm; 96 cm^2 7. 675 m^2 8. 16 m by 8 m

	π	Circumference	Area	Radius
9.	π	10π cm	25π cm^2	5 cm
10.	π	6π m	9π m^2	3 m
11.	π	5π cm	$6·25\pi$ cm^2	2·5 cm
12.	$\frac{22}{7}$	264 cm	5,544 cm^2	42 cm
13.	$\frac{22}{7}$	88 m	616 m^2	14 m
14.	3·14	157 mm	1,962·5 mm^2	25 mm
15.	3·14	125·6 m	1,256 m^2	20 m
16.	π	11π cm	$30·25\pi$ cm^2	5·5 cm
17.	$\frac{22}{7}$	66 m	346·5 m^2	10·5 m
18.	3·14	75·36 cm	452·16 cm^2	12 cm

19. 49 cm 20. 21 cm 21. (i) 41 cm (ii) 100 cm^2 22. 39π cm^2 23. 10 cm
24. 6 cm 25. 6·5 cm 26. 9 cm

Exercise 4.3

1. (i) 120 cm^3 (ii) 148 cm^2 2. (i) 576 m^3 (ii) 432 m^2 3. (i) 1,260 mm^3 (ii) 766 mm^2
4. (i) 5 cm (ii) 392 cm^2 5. 525 l 6. (i) 8 cm (ii) 860 cm^2 7. 576 8. 54 cm^2
9. 96 cm^2 10. 8 cm^3 11. 125 cm^3 12. 8 cm by 12 cm by 28 cm; 1,312 cm^2
13. (i) 9 cm (ii) 270 cm^3 14. (i) €12,000 (ii) 96 (iii) €25
15. (i) 1·5 cm (ii) 171 cm^2 (iii) 150

ANSWERS

Exercise 4.4
1. 400 cm³ 2. 8,160 cm³ 3. 864 cm³ 4. 400 cm³ 5. 1,350 cm³ 6. 3,840 cm³
7. 133·2 cm³ 8. 121 cm³ 9. 1,119 cm³ 10. (i) 240 cm² (ii) 96,000 cm³
11. 1·44 m³ 12. (i) 3,480 m³ (ii) 5 hours 48 minutes (iii) €41·76 13. (i) 6 cm² (ii) 20

Exercise 4.5

	π	Radius	Height	Volume	Curved Surface Area	Total Surface Area
1.	$\frac{22}{7}$	7 cm	12 cm	1,848 cm³	528 cm²	836 cm²
2.	3·14	15 cm	40 cm	28,260 cm³	3,768 cm²	5,181 cm²
3.	π	8 mm	11 mm	704π mm³	176π mm²	304π mm²
4.	$\frac{22}{7}$	3·5 m	10 m	385 m³	220 m²	297 m²
5.	3·14	12 cm	40 cm	18,086·4 cm³	3,014·4 cm²	3,918·72 cm²
6.	π	13 mm	30 mm	5,070π mm³	780π mm²	1,118π mm²

	π	Radius	Volume	Curved Surface Area
7.	$\frac{22}{7}$	21 cm	38,808 cm³	5,544 cm²
8.	3·14	9 m	3,052·08 m³	1,017·36 m²
9.	π	6 mm	288π mm³	144π mm²
10.	$\frac{22}{7}$	10·5 cm	4,851 cm³	1,386 cm²
11.	3·14	7·5 cm	1,766·25 cm³	706·5 cm²
12.	π	1·5 m	4·5π m³	9π m²

	π	Radius	Volume	Curved Surface Area	Total Surface Area
13.	π	15 mm	2,250π mm³	450π mm²	675π mm²
14.	π	$1\frac{1}{2}$ cm	$\frac{9}{4}\pi$ cm³	$\frac{9}{2}\pi$ cm²	$\frac{27}{4}\pi$ cm²
15.	$\frac{22}{7}$	42 cm	155,232 cm³	11,088 cm²	16,632 cm²
16.	3·14	12 m	3,617·28 m³	904·32 m²	1,356·48 m²

17. 24,492 cm³ 18. (i) 576 m² (ii) 3·2 m² (iii) 5% 19. 9
20. (i) 19,800 cm³ (ii) 4,851 cm³ (iii) 14,949 cm³

Exercise 4.6

	π	Radius	Height	Slant Height	Volume	Curved Surface Area
1.	π	8 cm	6 cm	10 cm	128π cm^3	80π cm^2
2.	$\frac{22}{7}$	21 mm	20 mm	29 mm	9,240 mm^3	1,914 mm^2
3.	3·14	3 cm	4 cm	5 cm	37·68 cm^3	47·1 cm^2
4.	π	1·5 m	2 m	2·5 m	$1·875\pi$ m^3	$3·75\pi$ m^2
5.	3·14	40 cm	9 cm	41 cm	15,072 cm^3	5,149·6 cm^2
6.	π	8 m	15 m	17 m	320π m^3	136π m^2
7.	$\frac{22}{7}$	2·8 cm	4·5 cm	5·3 cm	36·96 cm^3	46·64 cm^2
8.	π	4·8 mm	1·4 mm	5 mm	$10·752\pi$ mm^3	24π mm^2
9.	π	12 m	35 m	37 m	$1,680\pi$ m^3	444π m^2
10.	3·14	11 cm	60 cm	61 cm	7,598·8 cm^3	2,106·94 cm^2

11. (i) $123\frac{1}{5}$ cm^3 (ii) $316\frac{4}{5}$ cm^2 12. (i) 5,510·7 cm^3 (ii) 2,387·97 cm^2
13. (i) 720π cm^3 (ii) (a) 10 cm (b) $213\frac{1}{3}\pi$ cm^3 (c) $506\frac{2}{3}\pi$ cm^3
14. (i) 4 cm (ii) 234·5 cm^3 (iii) 135·59 cm^3
15. (i) 2·4 cm (ii) 42·24 cm^3 (iii) 617·76 cm^3
16. (i) πx^3 cm^3 (ii) 2 : 1 17. (i) 32π cm^3 (ii) 6 minutes

Exercise 4.7

2. 30π cm^3 3. (i) 936π m^3 (ii) 360π m^2 4. 4·1, 944π cm^3; 549π cm^2
5. (i) 795·048 cm^3 (ii) 431·436 cm^2 6. (i) 9,800 cm^2 (iii) 2,502,218 cm^3
7. (iii) 5·02 cm^3 (iv) 265   8. (i) 180 litres (ii) 268 litres (iii) 88 litres
9. (i) 15 m^2 (ii) 123·75 m^3 10. (i) 38,676 cm^3
11. (i) $350,000\pi$ m^3 (ii) $2,500\pi$ m^3 (iii) $\frac{5}{7}$%

Exercise 4.8

1. (i) 20 cm (ii) 240π cm^2 2. (i) 6 cm (ii) 288π cm^3 3. 15 cm
4. (i) 3 cm (ii) 36π cm^2 5. (i) 4 cm (ii) 80π cm^2 6. (i) 20 cm (ii) 1,570 cm^3
7. (i) $3\frac{1}{2}$ m (ii) 341 m^2 8. (i) 10 cm (ii) 96π cm^3 9. 10 cm
10. (i) 2 cm (ii) 35·2 cm^2 11. (i) 4 m (ii) 301·44 m^2 12. 45 cm
13. (i) 2·1 m (ii) $0·81\pi$ m^3 (iii) $0·096\pi$ m^3
14. (i) $\frac{250}{3}\pi$ cm^3 (ii) $\frac{25}{3}h\pi$ cm^3 (iii) 12 cm   15. (i) $\frac{3,087}{2}\pi$ cm^3 (ii) 42 cm
16. (i) $\frac{243}{4}\pi$ cm^3 (ii) 405π cm^3

ANSWERS

Exercise 4.9
1. $h = 9$ cm 2. $r = 2$ cm 3. $r = 6$ cm 4. $h = 18$ cm 5. $r = 5$ cm 6. $h = 7\frac{1}{2}$ cm
7. 20 cm 8. (i) $30{,}375\pi$ cm^2 (ii) 22·5 cm 9. 8 cm 10. (i) $\frac{32}{3}\pi$ cm^3
(ii) 150π cm^3 (iii) 0·43 cm 11. 400 12. (i) 3 cm (iv) 2 cm 13. (i) $121\cdot5\pi$ cm^3
(ii) 3·4 cm 14. (i) $300{,}000\pi$ cm^3 (ii) 28,125 (iii) 1 cm 15. (i) $l = 8$ m
(ii) 32 minutes 16. 20 cm 17. (i) $4\cdot5\pi$ cm^3 (ii) (a) 1·5 cm, 12 cm (b) 27π cm^3
(c) $\frac{2}{3}$ 18. (i) 6 (ii) 972 cm^3 (iii) 1·3 g 19. (i) 509 mm^3 (ii) (a) 48 mm (b) 18 mm
(c) 20 mm (iii) 17,280 mm^3 (iv) 5,064 mm^3 20. (i) 66 cm^3 (ii) 52 21. (i) $3{,}920\pi$ cm^3
(ii) 144π cm^3 (iii) 9 cm (iv) 27 22. (i) 1·75 m (iii) approximately 9 m^3

Exercise 4.10
1. 228 m^2 2. 750 m^2 3. 416 m^2 4. 464 m^2 5. 195 m^2 6. 265·5 m^2 7. 5 m
8. 5 cm 9. $\frac{47}{3}$ 10. 19·76; 79·04 m 11. (i) 3,600 m^2 (ii) 50%
12. (i) 12 cm (ii) (0, 5, 7, 4, 0); 48 cm^2 13. (i) 747 m^2 (ii) €28,386 14. (i) 74·75 cm^2
(ii) 78·5 cm^2 (iii) 3·75 cm^2 (v) 5% (vi) 76·2 cm^2 15. (i) 45·6 m^2 (ii) 285 m^2
16. (i) 27·4 m^2 (ii) (a) 822 m^3 (b) 822,000 l 17. (ii) 52·5 sq. units (iii) 1·6%
18. (i) $x = 3, y = 4\frac{1}{2}$ (ii) (a) 47 cm^2 (b) 23 km^2
19. Heights $\left\{7, 6\frac{1}{2}, 7\frac{1}{2}, 8, 5\frac{1}{2}, 0\right\}$ Area approx 93 cm^2

Exercise 4.11
1. (2, E), (3, A), (4, F), (5, B), (6, D)
2. (i)   (ii)   (iii)

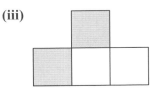

5. (i) Five faces (ii) Six vertices (iii) (a) 336 cm^2 (b) 288 cm^3 6. (i) Pyramid
(ii) Triangular prism 7. (ii) (a) 46,400 cm^2 (b) 4·64 m^2 8. (i) 26,400 cm^2
(ii) (a) 378,000 cm^3 (b) 0·378 m^3 (c) 378 l 10. (i) (a) Cuboid (b) Tetrahedron
(c) Cylinder (ii) (a) 160 cm^2 (b) $36\sqrt{3}$ cm^2 (c) 1,188 cm^2 (iii) (a) 144 cm^3 (c) 3,080 cm^3
11. (i) Cylinder (ii) $h = 8$ cm, $x = 62\cdot8$ cm 12. 4,060 cm^3

Exercise 5.1
1. 17, 21, 25, 29 2. 20, 15, 10, 5 3. −3, −1, 1, 3 4. 1, −2, −5, −8
5. 4·1, 4·5, 4·9, 5·3 6. 0·4, −0·2, −0·8, −1·4 7. 16, 32, 64, 128 8. 162, 486, 1,458, 4,374
9. 5, 7, 9, 11 10. 4, 7, 10, 13 11. 3, 7, 11, 15 12. 2, 7, 12, 17 13. −1, −3, −5, −7
14. −1, −5, −9, −13 15. 6, 9, 14, 21 16. 3, 8, 15, 24 17. 2, $\frac{3}{2}, \frac{4}{3}, \frac{5}{4}$ 18. 1, $\frac{4}{3}, \frac{3}{2}, \frac{8}{5}$
19. 2, 4, 8, 16 20. 3, 9, 27, 81 21. (i) 7, 12, 17 22. (i) 3, 6, 11 23. (i) $\frac{3}{2}, \frac{4}{3}, \frac{5}{4}$

Exercise 5.2

1. (ii) 1, 3, 5; Yes: odd numbers (iii) 2, 4, 6; Yes: even numbers (iv) Yellow (v) Green
   (vi) Green   2. (i) 1, 4, 7 (ii) 2, 5, 8 (iii) 3, 6, 9 (iv) Blue (v) White (vi) Red
3. (i) 1, 4, 7, 10 (ii) 2, 5, 8, 11 (iii) 3, 6, 9, 12 (iv) Yellow (v) Orange (vi) 23 (vii) 28
4. (i) Yellow: 1, 5, 9; Purple: 2, 6, 10; Green: 3, 7, 11; Blue: 4, 8, 12 (ii) Blue (iii) Purple
   (iv) Green (v) 31 (vi) 41

Exercise 5.3

1. 3, 5, 7, 9; 2   2. 5, 8, 11, 14; 3   3. 1, 5, 9, 13; 4   4. 7, 12, 17, 22; 5
5. 1, −1, −3, −5; −2   6. 5, 2, −1, −4; −3   7. 2, 5, 10, 17; 2   8. −1, 2, 7, 14; 2
9. 2, 6, 12, 20; 2   10. 2, 4, 8, 16   11. 3, 9, 27, 81   12. 5, 5, 11, 19

Exercise 5.4

1. 2, 4, 6, 8, 10; linear   2. 2, 5, 8, 11, 14; linear   3. 5, 9, 13, 17, 21; linear
4. 1, 4, 9, 16, 25; not linear   5. 0, 2, 6, 12, 20; not linear   6. 3, 8, 15, 24, 35; not linear
7. 2, 4, 8, 16, 32; not linear   8. 0, 4, 18, 48, 100; not linear   9. (i) 5, 7, 9, 11
10. (i) 1, 4, 7, 10, 13 (iv) 3   11. (i) Sequence B

12. (i)

Day	1	2	3	4	5
€	4	6	8	10	12

(ii) €22 (iii) €52 (v) €202 (vi) €18 (vii) 19 days

13. (i)

Day	1	2	3	...	10
€	1	4	7	...	28

(ii)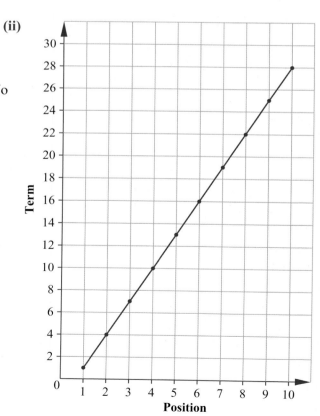

(iii) The values grow more quickly.
14. (ii) Amy (iv) linear (v) Day 18 (vi) No
15. (iii) Yes; day 21 (iv) Jane
16. (ii) Yellow flower (iii) Yes

ANSWERS

Exercise 5.5

1. $a = 1; d = 2; T_n = 2n - 1$ 2. $a = 2; d = 3; T_n = 3n - 1$ 3. $a = 3; d = 4; T_n = 4n - 1$
4. $a = 6; d = 5; T_n = 5n + 1$ 5. $a = 9; d = -2; T_n = 11 - 2n$ 6. $a = 4; d = -3; T_n = 7 - 3n$
7. $a = 8; d = -5; T_n = 13 - 5n$ 8. $a = 4; d = -6; T_n = 10 - 6n$ 9. $a = -5; d = 2; T_n = 2n - 7$
10. (i) $a = 1, d = 3$ (ii) $T_n = 3n - 2; T_{50} = 148$ (iii) T_{30} 11. (i) $a = 4, d = 5$
    (ii) $T_n = 5n - 1; T_{45} = 224$ (iii) T_{50} 12. (i) $a = 40, d = -4$ (ii) $T_n = 44 - 4n; T_{15} = -16$
    (iii) T_{11} 13. (i) $2n + 3$ (iii) $5n + 2$ (v) €55 14. (ii) $a = 5, d = 3$ (iii) (a) €2 (b) €3
    (iv) $32\frac{2}{3}$ km 15. (i) 35 mins (iii) $a = 35, d = 15$ (iv) $15n + 20$ (v) 155 mins (vi) $14\frac{2}{3}$ kg
16. T_{59} 17. T_{30} 18. (i) 28 months (ii) $2n + 8$ (iii) $3n + 5$ (iv) 338 months
19. $k = 3; 5, 7, 9$ 20. $k = 43; 3, 7, 11$ 21. $k = 11; 17, 23, 29, 35$ 22. (i) $k = 5; 3, 11, 19, 27$
    (ii) $T_n = 8n - 5; T_{21} = 163$ (iii) T_1

Exercise 5.6

1. (i) 9, 11, 13 (ii) arithmetic (iii) $2n + 1$ (iv) 101

2. (i)

Pattern	1	2	3	4	5
No. of tiles	7	13	19	25	31

   (ii) $6n + 1$ (iii) Pattern 42 3. (i) 17, 21 (ii) $4n + 1$ (iii) Shape 12
4. (i) 40 (ii) 10
5. (i)

Rectangle	1	2	3	4	5	6
Number of white squares	5	6	7	8	9	10
Number of red squares	4	6	8	10	12	14
Total number of squares	9	12	15	18	21	24

   (ii) 14 (iii) $n + 4$ (iv) $2n + 2$ (v) 22 (vi) 36 (vii) $3n + 6$
6. (i) 38 (ii) $7n + 3$ (iii) 143 (iv) $a = 4, b = 3$ (v) 9th diagram
7. (i) 4 cm (ii) 18 cm^2 (iii) 22 cm^2 (iv) 42 cm^2 (v) $(4n + 2)$ cm^2
8. (ii)

Diagram	1	2	3	4	5
Number of dots	5	8	11	14	17

   (iii) $3n + 2$ (v) Odd 9. (i) 1, 3, 6, 10 (iii) 0, 1, 3, 6 10. (i) 14, 20, 26 (iii) $6n + 8$
   (iv) Pond 15 (v) 4, 5, 6 (vi) $n + 3$ (vii) 18 slabs × 33 sabs

Exercise 5.7

1. $a = 3, d = 2$ 2. $a = 1, d = 3$ 3. $a = 7, d = 4$ 4. $a = 3, d = 5$ 5. $a = 5, d = 6$
6. (i) $d = 3$ (ii) $3n + 4; 64$ (iii) T_{24} 7. (i) $a = 3, d = 4$ (ii) $4n - 1$ (iv) $n = 25$
8. (i) $p = 7, q = 9$ (ii) 23 9. (i) $p = 10, q = 7, r = 1$ (ii) -47

Exercise 5.9
1. 120 2. 246 3. 603 4. (i) $\frac{n}{2}(5n + 1)$ (ii) 1,010 5. (i) $\frac{n}{2}(3n + 17)$ (ii) 1,605
6. (i) 5, 7, 9, 11 (ii) 2 (iii) 320 7. (i) $a = 2, d = 5$ 8. (i) $a = -7, d = 4$ (iii) (a) 69
(b) 1,530 9. (i) $a = 2, d = 3$ (iii) (a) 89 (b) 1,365 10. (i) $a = -15, d = 6$
(iii) (a) 129 (b) 0 11. (i) $a = 2, d = 6$ (ii) (a) 116 (b) 1,180 12. 610
13. 1,770 14. 6,275 15. (i) $3n - 1$ (iii) 620 16. (i) $10n$ (ii) $5n(n + 1)$ (iii) 201,000
17. (ii) (a) 39 (b) 400 (iii) 15 18. (i) $12a$ (ii) $75a$ (iii) $a = 2$ (iv) 6, 8, 10, 12
(v) (a) $2n + 2$ (b) $n(n + 5)$ (vi) (a) 42 (b) 500 19. (i) $a = 7, d = 2$ (ii) 10
20. (i) €10·20 (ii) €265 21. €492,000 22. 1,120 m 23. 90 24. (ii) €6,432
25. 30 seconds

Exercise 5.10
1. $a = 3, d = 2$ 2. $a = 4, d = 2$ 3. $a = 1, d = 6$ 4. $a = -1, d = 4$ 5. (i) 2 (ii) 23
6. (i) 3; 4 (ii) $4n - 1$ (iii) 39 7. (i) $a = -2, d = 4$ (ii) $4n - 6$ (iii) 74 (iv) 10

Exercise 5.11
1. $n^2 + n + 1$ 2. $n^2 + 2n + 1$ 3. $n^2 + n$ 4. $n^2 + 1$ 5. $n^2 - 3n + 2$ 6. $n^2 + 4n + 3$
7. (i) $n^2 + n - 2$ (ii) 108 (iii) Greater 8. (i) $n^2 + 11n + 10$ (ii) 220 m (iii) 4,270 m
9. (i) $n(n + 1)$ or $n^2 + n$ (ii) $2(n + 1)$

Exercise 6.1
1. (i) 10 (ii) 19 (iii) 1 (iv) 4 (v) 6 2. (i) 13 (ii) –12 (iii) –2 (iv) –7 (v) 0; 4
3. (i) 18 (ii) 4 (iii) 0 (iv) –2 (v) 12; –5, 2 4. (i) (a) 10 (b) 5 (ii) $k = 2$
5. $k = 3$ 6. (i) 3 (ii) –1 (iii) 8 (iv) –6; –5, 3 7. (i) –1 (ii) –2 (iii) $\frac{1}{4}$ or 0·25
(iv) $\frac{11}{56}$ or 0·56; 0, –2 8. (i) $\frac{1}{2}$ (ii) –1 (iii) 2; $x = 1$ 9. (i) 0·2 (ii) 0·5
(iii) 0·3125; $x = -4$; –1 10. (i) (a) 15 (b) 8 (c) 3 (ii) –4 (iii) ± 2
11. (i) 4; 2 (ii) 2 (iii) 1

Exercise 6.2
1. $k = 2$ 2. $h = 4$ 3. $a = 3$ 4. $b = -10$ 5. $a = 2$ 6. $k = -4$
7. $a = 2$ or $a = \frac{3}{2}$ 8. (i) $a = 2$ (ii) $b = -3$ 9. $a = 3; b = -4$ 10. $a = -7; b = -1$
11. (i) $a = 2; b = -3$ (ii) –1; 5 (iii) $-\frac{3}{2}$; 4 12. (i) $a = 5; b = -17$ (ii) 1, 2, 3, 4
13. $a + b = 1; a - b = 5; a = 3, b = -2$ 14. $p + q = 7; p - q = -3; p = 2, q = 5; -3, \frac{1}{2}$
15. (i) $a + b = -1; 2a + b = 1$ (ii) $a = 2, b = -3$ (iii) 0; 1 16. $c = -3, a = 4, b = -5$
17. (i) $q = -6$ (ii) –8; –3 18. $a = 2; b = -3$ 19. $b = 1, c = -2, k = -2$
20. $p = -3, q = -4; a = -1, b = 4, c = 6; \frac{3}{5}, 4$ 21. $a = 2, b = -5; h = -\frac{1}{2}, k = 3; -2, 3$
22. $b = -2, c = 4; p = 2, q = -1; -5, 1$ 23. $b = 3, c = -10; k = -10; 0, -3$
24. $-3; b = 2, C = -3$

ANSWERS

Exercise 7.1
16. (i) (2, −4)

Exercise 7.2
10. (i) (−1, 4), (2, −5)   (ii) $x = -1, x = 2$

Exercise 7.3
12. (−3, 0), (−2, 3), (1, 0)   13. (−1, 5), (3, −7)

Exercise 7.4
8. (i) $-\frac{1}{4}, -\frac{1}{2}, -2, 2, 1, \frac{1}{3}$ (ii) $x = 5$

Exercise 7.5
1. (i) −1, 2   (ii) max (0, 4); min (2, 0)   (iii) max value = 4, min value = 0
   (iv) 10·1   (v) $0 < x < 2$ (vi) $2 \le x < 0; 2 < x \le 4$ (vii) $-2 \le x < -1$ (viii) $0 \le k \le 4$
   (ix) (a) −0·7, 1, 2·7   (b) $-0.7 \le x \le 1; 2.7 \le x \le 4$ 2. (i) −3, −2, 1   (ii) $-3 \le x \le -2; 1 \le x \le 2$
   (iii) $-4 \le n \le -3; -2 \le x \le 0; 6$ 3. (i) −2, 1, 3   (ii) (a) $-2 < x < 1; 3 < x < 4$
   (b) $-3 < x < -2; 1 < x < 3$ (iii) −1·6, 0·2, 3·4   4. (iii) max (−1, 11); min (2, −16)
   (iv) (a) $-2.5 \le x \le -1; 2 < x < 3.5$ (b) $-1 < x < 2$ (c) $-2 < x < -1$ (v) −2, 0·3, 3·2
5. −1, 1, 2·5   (i) −3   (ii) −0·4, 0, 2·9   6. (i) (0, 5)   (ii) −1, 1·4, 3·6
   (iii) −0·7, 0·8, 3·9; $-20 \le h \le 4.5$ 7. (i) −1·8, 0, 3·3   (ii) −1·9, 0·2, 3·3   (iii) $3.3 \le x < 3.5$
   (iv) max (2, 20); min (−1, −7)   (v) (a) $-7 \le k \le 4$ (b) $4 < k \le 20$ 8. (i) −2, 2, 3
   (ii) −1, 0, 4   (iii) −1, 1, 3   (iv) $-1 \le x \le 1; 3 \le x \le 4$ (v) $-2 \le x \le -1; 1 \le x \le 3$
9. (i) 1, 3·5   (ii) $1 \le x \le 3.5$ (iii) $-2 \le x \le 1; 3.5 \le x \le 4$ 10. (i) −1, 0·5, 2   (ii) −1·1, 0, 2·6
   (iii) $-1.1 \le x \le 0; 2.6 \le x \le 3$ 11. (i) max (2, 6); min (0, 2)   (ii) 3·2
   (iii) $2 + 3x - x^3 = 2x + 2$ (v) 0, 1, 2   12. (iii) −3·7, −0·3, 2   (iv) −6
   (v) 6   (vi) −3, −1, 2   13. (i) −1, 4   (ii) −1, 2, 4   (iii) $x^3 - 5x^2 + 2x + 8 = 8 - 2x$
   (v) 0, 1, 4   (vi) $0 \le x \le 1; 4 \le x \le 5$ 14. $x(x^2 - 4)$; −2, 0, 2   (i) −2·4, 0·4, 2
   (ii) −2·2, 0, 2·2; 2·2   15. (i) max = 4; min = 0   (ii) max (−1, 4); min (1, 0)
   (iii) (a) $-1 < x < 1$ (b) $-3 \le x \le -2$ (c) $-2 < x < -1; 1 < x \le 3$ (iv) $0 < k < 4; -1.7, 0, 1.7; 1.7$
16. (iv) −1·6   17. 1·44

Exercise 7.6
1. (i) period = 8; range = [0, 6]   (ii) 6; 0; 3; 6   2. (i) period = 3; range = [0, 4]
   (ii) 4; 0; 4; 0   3. (i) period = 6; range = [1, 5]   4. (i) period = 8; range = [−3, 3]
5. (i) period = 20; range = [0, 4]   (iii) 40, 50, 60   (iv) (a) 75   (b) $60 \le x < 65; 70 \le x < 75$
6. (i) period = 12; range = [−8, 8]   (iii) 48, 52, 56, 60   (iv) (a) 42   (b) $38 < x < 42$
   (c) $36 < x < 38$ 7. $k = 2; a = -2, b = 3$

Exercise 7.7

1. (ii) $\frac{88}{3}$ 2. 78 3. 36 4. 32 5. 90 6. 54

Exercise 8.2

1. $4x^3$ 2. $6x^5$ 3. $6x$ 4. $-20x^3$ 5. 4 6. -3 7. 0 8. 0 9. $-3x^{-4}$ or $-\frac{3}{x^4}$
10. $-x^{-2}$ or $-\frac{1}{x^2}$ 11. $4x^3 + 6x^2$ 12. $6x^2 + 10x$ 13. $6x + 4$ 14. $4x - 6$ 15. $5 - 4x$
16. 3 17. $3x^2 + 4x + 5$ 18. $1 - 6x - 12x^2$ 19. $-10x - 24x^3$ 20. $2x - 1$
21. $3x^2 - 6x$ 22. $20 - 4x$ 23. $6x^2 - 16x + 7$ 24. $3x^2 - 4x + 4$ 25. $2 - 6x - 3x^2$
26. $3x^2 - 3x^{-4}$ or $3x^2 - \frac{3}{x^4}$ 27. $6x - 2x^{-3}$ or $6x - \frac{2}{x^3}$ 28. $4x^3 - 4x^{-5}$ or $4x^3 - \frac{4}{x^5}$

Exercise 8.3

1. 10 2. 43 3. 23 4. -5 5. 6 6. 13 7. -37 8. -20 9. 10π 10. 36π

Exercise 8.4

1. $12x + 22$ 2. $8x^3 + 22x$ 3. $6x^2 + 18x + 17$ 4. $5x^4 + 12x^3 - 9x^2 - 10x + 12$
5. $27x^2 - 42x + 19$ 6. $2x + 1$ 7. $-8x^3 + 9x^2 + 8x - 6$ 8. $4x^3 - 12x^2 - 5$
9. $3x^2 - 2x - 1$ 10. $-6x^5 - 5x^4 + 12x^3 + 15x^2 - 6$ 11. $6x^5 - 5x^4 + 4x^3 - 6x^2 - 4x$
12. $25x^4 - 60x^3 - 33x^2 + 36x + 6$ 13. $9x^2 - 8x - 4$ 14. $-5x^4 + 8x^3 + 2x - 2$
15. -2 16. -8 17. 7 18. -16

Exercise 8.5

1. $\dfrac{1}{(x+1)^2}$ 2. $\dfrac{5}{(x+3)^2}$ 3. $\dfrac{-1}{(x-1)^2}$ 4. $\dfrac{18}{(x+4)^2}$ 5. $\dfrac{-1}{(x+2)^2}$ 6. $\dfrac{-1}{(x-3)^2}$
7. $\dfrac{-3}{(x+4)^2}$ 8. $\dfrac{x^2 - 2x}{(x-1)^2}$ 9. $\dfrac{-23}{(5x-4)^2}$ 10. $\dfrac{-2x^2 + 6x + 2}{(x^2+1)^2}$ 11. $\dfrac{-4x}{(3x^2-1)^2}$
12. $\dfrac{12x}{(4-x^2)^2}$ 13. -8 14. 1 15. 4

Exercise 8.6

1. $5(2x + 3)^4(2)$ or $10(2x + 3)^4$ 2. $4(5x - 1)^3(5)$ or $20(5x - 1)^3$ 3. $3(x^2 + 3x)^2(2x + 3)$
4. $7(x^2 - 5x - 6)^6(2x - 5)$ 5. $6(4 - 5x)^5(-5)$ or $-30(4 - 5x)^5$ 6. $5(3 - 2x)^4(-2)$ or $-10(3 - 2x)^4$
7. $4(1 + x^2)^3(2x)$ or $8x(1 + x^2)^3$ 8. $7(5 - 2x^2)^6(-4x)$ or $-28x(5 - 2x^2)^6$
9. $8(4 - 3x - x^2)^7(-3 - 2x)$ 10. $5(x^2 + \frac{1}{x^2})^4(2x - 2x^{-3})$ or $5(x^2 + \frac{1}{x^2})^4(2x - \frac{2}{x^3})$
11. $4(x^3 - \frac{1}{x^3})^3(3x^2 + 3x^{-4})$ or $4(x^3 - \frac{1}{x^3})^3(3x^2 + \frac{3}{x^4})$ 12. $10(1 + \frac{1}{x})^9(-x^{-2})$ or $10(1 + \frac{1}{x})^9(\frac{-1}{x^2})$
13. 0 14. -30 15. 2 16. 24 17. -40 18. 0

ANSWERS

Exercise 8.7
1. −1   2. −7   3. 0   4. $3x - y - 6 = 0$ 5. $3x - y + 5 = 0$ 6. $6x - y + 5 = 0$
7. $x + y - 10 = 0$ 8. $3x - y + 1 = 0$ 9. $x + y - 2 = 0$; $x - y - 3 = 0$; yes

Exercise 8.8
1. (1, 1)   2. (−2, 13)   3. (4, −1)   4. (3, 0); (−1, −4)   5. (2, −20); (−1, 10)
6. (1, 5)   7. (3, −9); (−2, −4)   8. (0, 0); (−2, 2); $x - y = 0$; $x - y + 4 = 0$ 9. (i) 0   (iii) 3
(iv) −1, 2   10. −4   11. −3   12. (i) −2; 4   (ii) $20x - y + 28 = 0$; $20x - y - 80 = 0$

Exercise 8.9
1. (2, −1)   2. (−3, −8)   3. (2, −5)   4. (−3, 16)   5. (−2, 9)   6. (−2, 13)
7. max (−1, 9); min (3, −23)   8. max (1, 3); min (3, −1)   9. max (1, 6); min (2, 5)
10. max (1, 13); min (−3, −19)   11. max (−1, 10); min (−2, 9)   12. max (−2, 20); min (2, −12)
13. (ii) $3x^2 - 18x + 24$ (iii) max (2, 3); min (4, −1)   (v) $2 < x < 4$ 14. (ii) $a(-1, 0)$, $b(0, 25)$,
   $c(1, 32)$, $d(5, 0)$ 15. (i) $a = -3$; (2, −2)   (ii) (0, 2)   (iv) $-2 < k < 2$ 16. $2px + q$
17. $a = 2$, $b = -6$, $c = -3$

Exercise 8.10
1. (i) $x < 1$ (ii) $x > 1$ 2. (i) $x < -3$ (ii) $x > -3$

Exercise 8.11
1. (i) 15   (ii) 8   2. (i) 0   (ii) 12   3. (i) $3t^2 - 12t + 9$; $6t - 12$ (ii) 9 m/s   (iii) 6 m/s^2
(iv) $t = 1$ or $t = 3$ (v) $t = 2$ (vi) $t = 3$ (vii) $t = 5$ 4. (i) $3t^2 - 18t + 24$; $6t - 18$
(ii) 24 m/s   (iii) $t = 2$ or $t = 4$ (iv) 6 m/s   (v) $t = 3$ (vi) $t = 4$ 5. (i) 30 m
(ii) $18 - 3t$ (iii) 6 m/s   (iv) $t = 6$ (v) 54 m   (vi) −3 m/s^2 6. $\dfrac{d^2s}{dt^2} = -2$ (a constant)
(i) 10 m/s   (ii) 25 m   7. (i) 24 m   (ii) $16 - 4t$ (iii) 4 m/s   (iv) 14 m   (v) $t = 4$; 32 m
8. (i) $30 - 10t$ (ii) 15 m/s   (iii) −10 m/s^2 (iv) $t = 3$; 46 m   9. (i) 8 m/s^2 or −8 m/s^2
(ii) 18 m/s   10. (i) 9 m/s^2 or −9 m/s^2 (ii) 18·75 m/s   11. (i) $8 - 2t$ (ii) −4
(iii) $t = 4$ (iv) (a) 6   (b) 3   12. (i) $-20p^{-2}$ or $-\dfrac{20}{p^2}$ (ii) $-\dfrac{1}{5}$

Acknowledgments

The authors would like to thank Colman Humphrey, Elaine Guildea and Sorcha Forde, who helped with proofreading, checking the answers and making many valuable suggestions that are included in the final text.

The authors also wish to express their thanks to the staff of Gill & Macmillan, and special thanks to Tess Tattersall, for their advice, guidance and untiring assistance in the preparation and presentation of the book.